Introduction to Reliability and Quality Engineering

Second edition

To Pauline

Introduction to Reliability and Quality Engineering

Second edition

DR JOHN BENTLEY BSc PhD CEng FInst MC

Principal Lecturer in Instrumentation
School of Science and Technology
The University of Teesside

Harlow, England • London • New York • Boston • San Francisco • Toronto • Sydney • Singapore • Hong Kong
Tokyo • Seoul • Taipei • New Delhi • Cape Town • Madrid • Mexico City • Amsterdam • Munich • Paris • Milan

Pearson Education Limited
Edinburgh Gate
Harlow
Essex CM20 2JE
England

and Associated Companies throughout the world

Visit us on the World Wide Web at:
http://www.pearsoneduc.com

First published 1993
Second edition 1999

Text design by Claire Brodmann
Typeset by 35 in 10/12 pt Times
Printed and bound in Great Britain by Antony Rowe Ltd, Chippenham and Eastbourne

ISBN 0201-33132-2

British Library Cataloguing-in-Publication Data
A catalogue record for this book is available from the British Library

Library of Congress Cataloguing-in-Publication Data
Bentley, John P., 1943–
    Introduction to reliability and quality engineering / John
Bentley. — 2nd ed.
        p.   cm.
    Includes bibliographical references and index.
    ISBN 0–201–33132–2
    1. Reliability (Engineering)   2. Quality control.   I. Title.
TS173.B45   1999
620′.0045—dc21   98–34263
                                                  CIP

10 9 8 7 6 5 4
07 06 05 04

Contents

Preface to the second edition

The quality of an engineering product is the set of characteristics which determine customer satisfaction; these define the product specification. In a highly competitive environment, a customer will only buy a product if it has an appropriate quality level or specification. Furthermore, if once the product is purchased, it does not meet the specification, then customer dissatisfaction will limit further sales. The survival of any manufacturing unit, therefore, depends critically on the quality of its products. During the 1970s the quality of UK engineering products was much criticized but since then many firms have implemented Total Quality Management with corresponding product improvements. The reliability of a product is its ability to retain its quality, that is, to continue to meet the specification, over a given period of time. Reliability is therefore the projection of initial quality into future time; if a product prematurely falls outside its specification, it will cause economic loss to the customer and possibly present a safety hazard. Quality and Reliability are therefore interrelated and together are vital in any engineering activity; thus it is essential that their joint study forms part of any engineering course.

This book is an introduction to the topics of quality and reliability, using a broad approach which is not linked to any particular engineering discipline.

The second edition has been expanded to include more material on statistical quality/process control, the economics of quality/reliability and fault tolerant systems for improving reliability. More worked examples have been included both to broaden the coverage of topics and to demonstrate theory in practice. These are supported by a full range of self-assessment questions.

Chapter 1 begins by explaining the meaning of quality and then discusses statistical techniques for quantifying and controlling quality. Chapter 2 defines the concept of reliability and then shows how it is quantified using practical definitions and statistical distributions. Chapter 3 shows how to calculate the reliability of a range of important systems, given failure rate data for individual elements and components. Typical component and element failure rate data and models are presented in Chapter 4, together with discussion of software and human reliability. Chapter 5 deals with quality and reliability in design and manufacture. It first discusses Total Quality Management, the influence of quality and reliability on product economics and how both can be achieved at the design, manufacture and testing stages. Chapter 6 is concerned with reliability and availability in maintenance and examines how different maintenance strategies influence the total cost of ownership; these strategies are then discussed in detail. The application of reliability principles to engineering safety is covered in Chapter 7: this is a case study of the design of a highly reliable protective system for a hazardous chemical process.

The book has around 100 line diagrams and tables and 17 worked examples positioned at key points. There are self-assessment questions at the end of each chapter with a total of 42. The mathematics is limited to the minimum necessary for understanding the principles involved.

The book is intended primarily for undergraduate students on electrical, electronic, instrumentation/control, mechanical, manufacturing and chemical engineering degree courses. Furthermore, since the level of mathematics is not high, much of the material will be helpful to technician engineers and students. The book should also be of use to professional engineers requiring an introductory text.

John P. Bentley
Guisborough 1998

Acknowledgements

The author would like to thank his wife Pauline for her constant support and her patient typing of the manuscript.

We are indebted to the following for permission to reproduce copyright material:

Hodder & Stoughton Ltd/New English Library Ltd and the author, M. Haslehurst, for Figure 5.19 (Haslehurst, 1969) Copyright © 1969 M. Haslehurst; Institute of Measurement & Control for Table 4.8 (Hellyer, 1985); Institute of Physics Publishing Ltd for Tables 4.2, 4.3, and 4.9 (Wright, 1984); The Institution of Chemical Engineers and the author, R. M. Stewart, for Figures 7.2, 7.8–7.12, and Table 7.3 (Stewart, 1971); the author, Professor F. P. Lees, for Tables 4.5–4.7 and 6.1 (Lees, 1976); Macmillan Accounts and Administration Ltd for Figure 4.3 (Carter, 1986) and Table 4.4 (Smith, 1988).

While every effort has been made to trace the owners of copyright material, in a few cases this has been impossible and we take this opportunity to offer our apologies to any copyright holders whose rights we may have unwittingly infringed.

Symbols and abbreviations

A	availability
ADC	analogue-to-digital converter
ATE	automatic test equipment
AQL	acceptable quality level
α	producer's risk
β	Weibull shape parameter, customer's risk
bcd	binary coded decimal
c	limit on number of defects, loss function constant
C_I	user initial cost
C_L	repair labour cost per hour
C_M	average materials cost per service
C_{PB}	process cost per hour following a breakdown
C_{PP}	process cost per hour during servicing
C_R	average materials cost per repair
C_P	potential capability index
C_{PK}	actual capability index
C_T	total manufacturing cost
c_0	fixed manufacturing cost
c_Q	manufacturing cost per unit
c_λ	manufacturer's cost per failure
CMS	common mode system
DR	demand rate
$\delta\lambda$	tolerance limit on λ
ΔT_W	extension of wear-out time
E	activation energy
E_T	total number of errors
ETA	Event tree analysis
$\varepsilon_R(\tau)$	fractional number of residual bugs at τ

$\varepsilon_C(\tau)$	fractional number of corrected bugs at τ
F, $F(t)$	unreliability
F_D	fail-danger probability
F_S	fail-safe probability
FAR	fatal accident rate
FDT	fractional dead time
FMEA	failure mode and effect analysis
FS	functional system
FTA	fault tree analysis
H	hazard rate
HIPS	high integrity protective system
HISS	high integrity shut down system
HITI	high integrity trip initiator
HIVE	high integrity voting equipment
I_T	total number of instructions
i,j	given sample, item, failure, term
k	Boltzmann's constant, term
K	arbitrary constant
$l(x)$	user's loss function
L	performance statistic
LED	light emitting diode
LTPD	lot tolerance per cent defective
LCL	lower control limit
LSL	lower specification limit
$\{LSL\}$	set of lower specification limits
λ	failure rate, assumed constant with time
$\lambda(t)$	instantaneous failure rate, hazard rate
$\bar{\lambda}$	mean failure rate, observed failure rate
λ_P	predicted failure rate
λ_T	target failure rate
m	maintenance frequency
m oo n	m out of n voting
MDT	mean down time
MMT	mean maintenance time
MTBF	mean time between failures
MTTF	mean time to fail
MTTR	mean time to repair
MTTT	mean time to test
μ	repair rate, mean value
N, n	number of values, items, samples
N_F	number of failures
η	Weibull scale parameter
OCC	operating characteristic curves

p	probability density, design parameter
$\{p\}$	set of design parameters
$p(x)$	probability density function
P	probability, profit
PFI	potential failure interval
π_E	environmental factor in failure rate
π_Q	quality factor in failure rate
π_T	temperature factor in failure rate
r	random variable, binomial parameter
$\{r\}$	set of random variables
R	range
$\bar{R}$	mean of sub-group ranges
R_j	sub-group range
$R, R(t)$	reliability
ρ	fractional rate of removing errors
s	selling price per item, standard deviation
S_T	total sales income
S	system
S/H	sample and hold device
SM	safety margin
SPC	statistical process control
SQC	statistical quality control
σ	standard deviation
$\{\sigma\}$	set of standard deviation values, vector
t	time
T	test interval, system lifetime, trip testing interval
T_C	condition monitoring interval
T_{Dj}	**down** time associated with jth failure
T_i	survival time or **up** time for ith failure
T_M	maintenance or service interval, testing interval
$\bar{\bar{T}}_W$	mean wear-out time
TQM	total quality management
τ	normalized time, debugging time
UCL	upper control limit
USL	upper specification limit
$\{USL\}$	set of upper specification limits
U	unavailability
x	continuous performance characteristic
x_T	target value of x
$\{x\}$	set of values of x
$\{x_T\}$	set of values of x_T, target vector
$\bar{x}$	mean value of x
x_{RMS}	root mean square value of x

X	trip initiating parameter
X_{T}	trip setting
y	discrete performance characteristic
z	binary performance characteristic, Poisson distribution parameter
ξ	time
t_{o}	Weibull time origin

1

Principles of quality

1.1 Introduction

The word **quality** is commonly used in everyday life. Manufacturers' advertisements constantly claim the quality of their products. Consumers organizations and magazines assess the quality of products such as cars, washing machines and computers. There is much debate about the quality of services in the water and gas industries, rail transport and the health and education services. This chapter begins by explaining the meaning of quality and then discusses statistical techniques for quantifying and controlling quality.

1.2 The quality of a product

1.2.1 The meaning of quality

The quality of a product or service can be defined as its ability to ensure complete customer satisfaction. However, consumer satisfaction with a given product or service will be based on **more than one feature**. Thus the quality of a meal in a restaurant will be judged on the comfort of the surroundings, degree of choice offered by the menu, efficiency of the service as well as the quality of the meal itself. Thus a more appropriate definition of quality is:[1] 'the totality of features and characteristics of a product, process or service that bear on its ability to satisfy stated or implied needs'.

The quality of an engineering product can therefore be measured in terms of a number of characteristics that contribute to an overall performance which satisfies a customer's requirements; these are termed **performance characteristics**.

Table 1.1   Different types of performance characteristics for a family car

Continuous {x}	Discrete {y}	Binary {z}
Urban fuel consumption	Visual appeal of body style	Leaded/unleaded petrol?
Maximum speed	Visual appeal of interior	Starts first time?
Time from 0 to 60 m.p.h.	Comfort of ride	Central locking?
Braking distance at 60 m.p.h.	Range of exterior colours	Quad stereo?
Engine noise level	Range of interior colours	Power-assisted steering?
% CO in exhaust		Tinted glass?
		Sun-roof?

Table 1.1 lists the main performance characteristics for a family car. The table shows that the characteristics can be divided into three types: **continuous**, **discrete** and **binary**.

Continuous characteristics can take any value within a given range. Thus, fuel consumption can take any value between 10 and 60 miles per gallon, while the time required to accelerate from 0 to 60 miles per hour may be any value between 8 and 20 seconds. These continuous variables are also **objective**; their values can be accurately established by independent measurement and are not dependent on the opinion of any individual customer. Continuous characteristics are also referred to as **variables**.

Discrete characteristics are used for the more difficult problem of assessing less tangible characteristics of the car; examples are the visual appeal of the body style or the interior. It will normally be possible to allocate only between four and six discrete values to such a characteristic; for example, the visual appeal of the body style could be assessed using the five numbers 5 (Excellent), 4 (Good), 3 (Satisfactory), 2 (Unsatisfactory), 1 (Repulsive!). These discrete characteristics are also **subjective**, that is, dependent on the opinion of an individual customer. Thus one customer may regard a given body style as excellent, whereas another may judge it to be merely satisfactory.

The third type of characteristic is termed **binary**. This refers to features that the car either does or does not possess, or tests that the car can either pass or fail; therefore binary characteristics can take only two values 0 or 1 corresponding to present/not present or pass/fail. Examples are ability/inability to run on unleaded petrol, ability/inability to start first time, presence/absence of sun-roof, presence/absence of tinted window glass. Binary characteristics are also referred to as **attributes**.

Thus summarizing, the quality of an engineering product can be specified by three sets of output performance characteristics (Figure 1.1). The set of numbers $x_1, x_2, \ldots$, *i.e.* the vector $\{x\}$, specifies the continuous characteristics; the set of numbers $y_1, y_2, \ldots$, *i.e.* the vector $\{y\}$, specifies the discrete characteristics; the set of numbers $z_1, z_2, \ldots$, *i.e.* the vector $\{z\}$, specifies the binary characteristics.

1.2.2 Product specification using continuous performance characteristics

Of the three types of characteristics discussed above, continuous characteristics are the most useful in quantifying product performance. Unlike discrete characteristics,

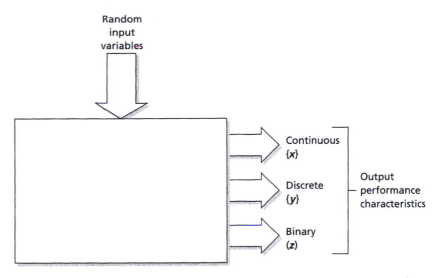

Figure 1.1 Performance characteristics of an engineering product

continuous characteristics are objective in that their value can be established by independent measurement. Unlike binary characteristics, continuous characteristics can indicate small changes in quality. The manufacturer of an engineering product will need to produce a **specification** which defines the product for a potential customer. This specification is best stated in terms of continuous variables and should consist of a set of **target values** $x_{1T}, x_{2T}, \ldots$, *i.e.* a target vector $\{x_T\}$. Thus the specification for a family car could consist of the following set of target values:

Urban fuel consumption:	40 miles per gallon
Maximum speed:	100 miles per hour
Time to 60 m.p.h.:	13 seconds
Braking distance at 60 m.p.h.:	180 feet
Engine noise level:	70 dB
Carbon monoxide in engine exhaust:	1 %

However, all engineering products are subject to **random effects** in the components which make up the product, in the manufacturing process and in the environment in which the final product is placed. If we consider a batch of 'identical' metallic engine components, then individual components in the batch will exhibit small random variations in dimensions, surface finish and chemical composition. Similarly, individuals in a batch of 'identical' tyres or brake pads will show small random variations in surface condition. The environment in which the car is placed will depend on geographic location and weather conditions; there will be corresponding random variations in ambient temperature, pressure, humidity, salinity, light level and road surface conditions. These random effects can be regarded as **random inputs** to the product which result in corresponding random variations in the output continuous performance characteristics $\{x\}$

of the product (Figure 1.1). These variations take place around the target values $\{x_T\}$. Thus the braking distance of individual cars in a batch of a given model, at 60 m.p.h., on a good dry road surface, may vary between 179 and 181 feet due to random variations in brake pad and tyre conditions, *etc.* When wet and icy road conditions are taken into account the variation is much wider.

The above random deviations in $\{x\}$ from target values $\{x_T\}$ will be a source of dissatisfaction to the customer. If a customer buys an individual car with an urban fuel consumption of 36 m.p.g., which is below the model target figure of 40 m.p.g., then he faces higher petrol costs. If the customer buys an individual car with an above-target fuel consumption of 44 m.p.g., then the petrol costs are reduced. This improved fuel consumption may, however, correspond to a worsening of other performance characteristics such as acceleration or ease of starting. The manufacturer is therefore compelled to place limits on the deviations in $\{x\}$, for individual products, from the target values $\{x_T\}$; these are called **specification limits**. If we consider the performance characteristic x_1 with target value x_{1T}, the manufacturer will reject for sale all individual products which fall outside the range $x_1 = LSL_1$ to $x_1 = USL_1$, where LSL_1, USL_1 are the lower and upper specification limits associated with x_1. Thus if a manufacturer of carbon resistors states that a resistor with a 10 kΩ target value has ±5% specification limits, then all resistors with resistance less than LSL = 9.5 kΩ and resistance greater than USL = 10.5 kΩ must be rejected for sale. Extending the argument to the full set of performance characteristics, a product should be rejected for sale if x_1 falls outside LSL_1 to USL_1, or x_2 falls outside LSL_2 to USL_2, or x_3 falls outside LSL_3 to USL_3, *i.e.* if any element of $\{x\}$ falls outside the corresponding elements of $\{LSL\}$, $\{USL\}$. Thus the product specification can be stated in terms of three vectors, the target vector $\{x_T\}$, and the specification limit vectors $\{LSL\}$, $\{USL\}$.

1.2.3 Statistical analysis of continuous performance characteristics

Both manufacturer and customer will need to know how the actual random variations in a given performance characteristic x, for individuals of a given product, compare with the target value x_T and the specification limits LSL, USL. A statistical analysis of these variations is required; this involves estimating the **mean, standard deviation, probability** and **probability density function** of the variations. To do this we take N **sample** values from the total **population** of values of x. These are specified by x_i where $i = 1, 2, \ldots, N$.

Mean $\bar{x}$

This specifies the centre of the spread of the sample values and is given by:

$$\bar{x} = \frac{1}{N} \sum_{i=1}^{i=N} x_i \qquad \textbf{mean}$$

(1.1)

Standard deviation

This specifies the average spread or deviation of the sample values from the mean value and is given by:

$$s = \sqrt{\frac{1}{N}\sum_{i=1}^{i=N}(x_i - \bar{x})^2} \qquad \textbf{standard deviation} \tag{1.2}$$

A related quantity to standard deviation is **root mean square** (RMS) value x_{RMS}, where:

$$x_{RMS} = \sqrt{\frac{1}{N}\sum_{i=1}^{i=N}x_i^2} \qquad \textbf{root mean square value} \tag{1.3}$$

We note that in the special case that mean $\bar{x} = 0$, standard deviation s is equal to root mean square value x_{RMS}.

Probability and probability density

The mean and standard deviation specify the centre and spread of the variation in the sample values, but give no information as to the **shape** of the variation, *e.g.* whether the sample values are concentrated around the centre or the extremes of the variation. This information can be obtained from probability and probability density distributions.

The term **probability** is used to refer to the likelihood of a particular event occurring. A probability of 0 is taken to represent an absolute impossibility and a probability of 1 an absolute certainty, thus all probability values lie between 0 and 1. If a large number of random, independent trials are made, then the probability P of a particular event occurring is given by the ratio:

$$P = \frac{\text{Number of occurrences of the event}}{\text{Total number of trials}} \tag{1.4}$$

in the limit that the total number of trials tends to infinity. Thus the probability of a tossed coin showing heads tends to the theoretical value of 0.5 over a large number of trials.

We can now calculate the probability that the performance characteristic x lies within a certain range of values. Figure 1.2(a) shows the horizontal x axis divided into m sections each of width δx. We then count the number of samples occurring in each section, *i.e.* n_1 in Section 1, n_2 in Section 2, n_j in Section j, *etc.*, where $j = 1, \ldots , m$. From Equation 1.4, the **probability** P_j of the characteristic occurring in the jth section is:

$$P_j = \frac{\text{Number of times sample occurs in the } j\text{th section}}{\text{Total number of samples}}$$

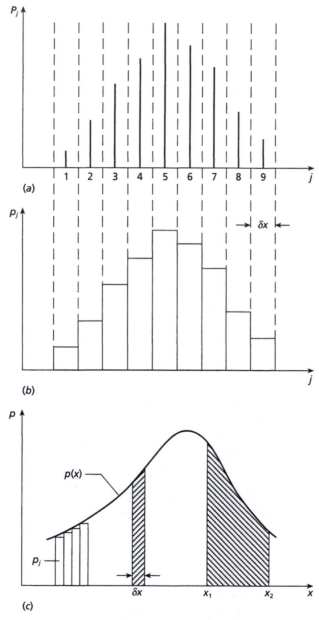

Figure 1.2 Probability distributions: (a) discrete probability; (b) discrete probability density; (c) continuous probability density

i.e.:

$$P_j = \frac{n_j}{N} \quad j = 1, 2, \ldots, m \tag{1.5}$$

In Figure 1.2(a) the height of each vertical line is equal to probability P_j; this is a **discrete probability distribution**.

A more useful distribution is that of **probability density**; probability density values p_j are obtained by dividing the above probability values P_j by the interval δx, *i.e.*:

$$p_j = \frac{P_j}{\delta x} \tag{1.6}$$

i.e. $P_j = p_j \delta x$. Figure 1.2(b) is a **discrete probability density distribution** or histogram. It consists of a series of rectangles of height p_j and width δx; the area of each rectangle is $p_j \delta x = P_j$. The **area of each rectangle** therefore is equal to the probability P_j of the characteristic occurring in the jth section. The sum of the areas of all the rectangles corresponds to the total probability of x taking any value; this is equal to 1.

Figure 1.2(c) shows that in the limit the interval δx tends to zero, and the discrete probability density distribution p_j tends to a continuous function.

This is the **probability density function $p(x)$** which is defined by:

$$p(x) = \lim_{\delta x \to 0} p_j \tag{1.7}$$

From Equation 1.6, the probability $P_{x,x+\delta x}$ that the characteristic will lie between x and $x + \delta x$ is given by:

$$P_{x,x+\delta x} = p(x)\,\delta x \tag{1.8}$$

i.e. by the area of the strip of height $p(x)$ and width δx (Figure 1.2(c)). Similarly the probability P_{x_1,x_2} that the characteristic will lie between x_1 and x_2 is given by:

$$P_{x_1,x_2} = \int_{x_1}^{x_2} p(x)\,\mathrm{d}x \tag{1.9}$$

i.e. by the area under the probability density curve between x_1 and x_2. The total area under the probability density curve is equal to 1, corresponding to the total probability of the characteristic having any value of x.

Normal or Gaussian probability density function

Statistical variations in performance characteristic x due to random effects in components, manufacturing and the environment can often be satisfactorily described by a **Gaussian or normal probability density function**. This has the equation:

$$p(x) = \frac{1}{\sigma\sqrt{(2\pi)}} \exp[-(x-\mu)^2/2\sigma^2] \qquad \text{Gaussian probability density function} \qquad (1.10)$$

where μ = mean value (specifies centre of distribution)

σ = standard deviation (specifies spread of distribution)

By making the substitution:

$$z = \frac{x - \mu}{\sigma}$$

The normal distribution can be written in the standard form:

$$p(z) = \frac{1}{\sqrt{2\pi}} \exp\left(\frac{-z^2}{2}\right) \qquad (1.11)$$

Values of $p(z)$ are given in Table 1.2 and Figure 1.3 shows the form of the distribution. We note that the distribution is symmetric about the origin $z = 0$, i.e.

$$p(-z) = p(z)$$

Also $p(z)$ is maximum when $z = 0$, which corresponds to $x = \mu$.

Figure 1.3 also shows the shaded area under $p(z)$ between $z = u$ and $z = \infty$. This gives the probability $P_{u,\infty}$ of z lying between u and ∞, *i.e.* being greater than u. Values of $P_{u,\infty}$ are calculated using:

$$P_{u,\infty} = \int_u^\infty p(z)\,dz = \frac{1}{\sqrt{2\pi}} \int_u^\infty \exp\left(\frac{-z^2}{2}\right) dz \qquad (1.12)$$

and are given in Table 1.2. We can use these values and the properties.:

(a) $P_{u,\infty} = P_{-\infty,u}$ (curve is symmetrical about $z = 0$)
(b) $P_{-\infty,+\infty} = 1$ (total area under curve is 1)

to calculate other required probabilities. Important results are:

(i) $P_{-1,+1} = P_{\mu-\sigma,\,\mu+\sigma} = 0.683$
(ii) $P_{-2,+2} = P_{\mu-2\sigma,\,\mu+2\sigma} = 0.955$
(iii) $P_{-3,+3} = P_{\mu-3\sigma,\,\mu+3\sigma} = 0.997$

which represent the probabilities of x lying between $\pm1, \pm2, \pm3$, standard deviations from the mean respectively.

Table 1.2 Values and areas of standardized Gaussian probability density function

z, u	p(z)	$P_{u,\infty}$
0.0	0.3989	0.5000
0.5	0.3521	0.3085
1.0	0.2420	0.1587
1.5	0.1295	0.0668
2.0	0.0540	0.0228
2.5	0.0175	0.0062
3.0	0.0044	0.0014
3.5	0.0009	0.0002
4.0	0.0001	0.0000

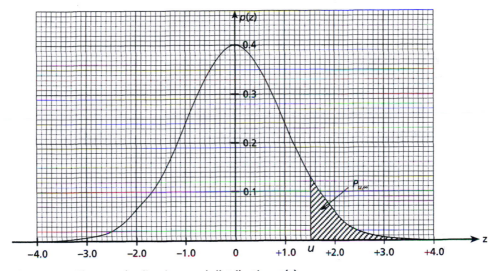

Figure 1.3 The standardized normal distribution $p(z)$

Example 1.1

A normal probability density function $p(x)$ has a mean value $\mu = 100$ and standard deviation $\sigma = 10$. Calculate the probability of x:

(a) being greater than 120
(b) being less than 80
(c) lying between 80 and 120
(d) being greater than 105
(e) being less than 95
(f) lying between 95 and 105
(g) lying between 90 and 130.

Solutions

(a) when $x = 120$, $u = \dfrac{x - \mu}{\sigma} = \dfrac{120 - 100}{10} = 2.0$, $P_{2.0,\infty} = 0.0228$

$\mathbf{P_{x>120} = 0.0228}$

(b) when $x = 80$, $u = \dfrac{x - \mu}{\sigma} = \dfrac{80 - 120}{10} = -2.0$, $P_{-\infty,-2.0} = P_{2.0,\infty} = 0.0228$

$\mathbf{P_{x<80} = 0.0228}$

(c) $P_{-\infty,+\infty} = 1$

$P_{-\infty,-2.0} + P_{-2.0,+2.0} + P_{2.0,\infty} = 1$

$0.0228 + P_{-2.0,+2.0} + 0.0228 = 1$

$P_{-2.0,+2.0} = 1 - 0.0456 = 0.9544$

$\mathbf{P_{80,120} = 0.9544}$

(d) when $x = 105$, $u = \dfrac{105 - 100}{10} = 0.5$, $P_{0.5,\infty} = 0.3085$

$\mathbf{P_{x>105} = 0.3085}$

(e) when $x = 95$, $u = \dfrac{95 - 100}{10} = -0.5$, $P_{-\infty,-0.5} = P_{0.5,\infty} = 0.3085$

$\mathbf{P_{x<95} = 0.3085}$

(f) $P_{-\infty,-0.5} + P_{-0.5,+0.5} + P_{+0.5,+\infty} = 1$

$0.3085 + P_{-0.5,\,0.5} + 0.3085 = 1$

$P_{-0.5,0.5} = 1 - 0.617 = \mathbf{0.383}$

$\mathbf{P_{95,105} = 0.383}$

(g) when $x = 90$, $u = \dfrac{x - \mu}{\sigma} = \dfrac{90 - 100}{10} = -1.0$

$P_{-\infty,-1.0} = P_{1.0,\infty} = 0.1587$

when $x = 130$, $u = \dfrac{x - \mu}{\sigma} = \dfrac{130 - 100}{10} = +3.0$

$P_{3.0,\infty} = 0.0014$

$P_{-\infty,-1.0} + P_{-1.0,+3.0} + P_{3.0,\infty} = 1.0$

$0.1587 + P_{-1.0,3.0} + 0.0014 = 1.0$

$P_{-1.0,3.0} = 0.8399$

$\mathbf{P_{90,130} = 0.8399}$

1.3 Statistical quality control, statistical process control

We saw in the previous section that the quality of a product can be measured using performance characteristics. Continuous characteristics (or variables) of the product exhibit random variations which can be represented by statistical distributions. The quality of the product is defined using a target value and upper/lower specification limits for each characteristic. We can then compare the statistical distribution of each characteristic with these specification limits in order to decide whether the product should be accepted or rejected for sale. This is called **statistical quality control (SQC)**. There are two main sources of these variations in product characteristcs:

(a)   The variability of bought-in components and materials
(b)   The variability of the production process itself.

Thus the variability of the molten iron product from a blast furnace is caused by variations in the iron ore and coke feedstock and by variations in temperature, heat and material flow rates inside the furnace. SQC techniques can be used to identify and limit both sources of variability. When applied to processes SQC is referred to as **statistical process control (SPC)**. Here online measurements of product performance characteristics are made and compared with the specification limits. This will not only inform the operators of the percentage of off-specification product but should also identify sources of variability in the process that can be reduced.

1.3.1 Quality control based on individuals

This is the simplest quality control strategy and is shown in Figure 1.4. A given continuous performance characteristic x is measured for all items of a given product

$$
\begin{array}{ll}
\text{If} \quad \text{LSL} \leqslant x \leqslant \text{USL} & \text{Accept} \\
\text{If} \quad \text{LSL} > x > \text{USL} & \text{Reject}
\end{array}
\tag{1.13}
$$

where LSL, USL are the lower and upper specification limits respectively.

Figure 1.5 shows the probability density function for a continuous performance characteristic x; μ is the mean value of x for the total population of products. The target value x_T and lower and upper specification limits LSL, USL are also shown. The probability of the product meeting the specification is given by $P_{\text{LSL,USL}}$, the area under the distribution between $x = \text{LSL}$ and $x = \text{USL}$. The probability of the product not meeting the specification is given by $P_{x<\text{LSL}} + P_{x>\text{USL}}$, *i.e.* the sum of the shaded areas. As an example, we consider a normal probability density function which has a mean $\mu = $ target value, x_T. If USL $= \mu + 3\sigma$, LSL $= \mu - 3\sigma$ then the total probability of the product failing to meet the specification is equal to $P_{x<\mu-3\sigma} + P_{x>\mu+3\sigma}$ From Section 1.2.3 the values of z corresponding to $x = \mu - 3\sigma$, $x = \mu + 3\sigma$ are $z = -3$ and $z = +3$, and the total probability of the product not meeting the specification is: $P_{-\infty,-3} + P_{+3,+\infty} = 2P_{3,\infty} = 0.0028$ (Table 1.2).

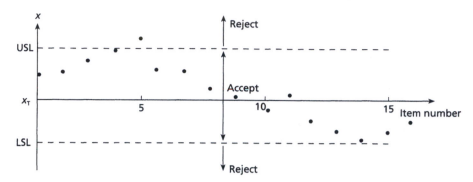

Figure 1.4 Quality Control based on individuals

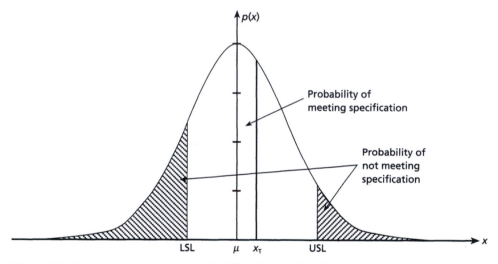

Figure 1.5 Comparison between probability density function and specification

1.3.2 Process capability indices

It is essential to know whether a given process is capable of making a product which meets the specification for that product. **Capability indices** compare the standard deviation σ of a product characteristic, caused by process variability, with the product specification limits LSL, USL.

Potential capability index C_P is a measure of the theoretical or inherent capability of a process to meet the product specification. It is defined by:

$$C_P = \frac{\text{USL} - \text{LSL}}{6\sigma}$$

(1.14)

C_P is used at the process design stage:

If $C_P = 1.0$, process is just potentially capable of meeting the specification.
If $C_P < 1.0$, process is potentially incapable.
If $C_P > 1.0$, process is potentially capable.

Thus if USL $= \mu + 4\sigma$, LSL $= \mu - 4\sigma$, $C_P = 8/6 = 1.33$ and the process is potentially capable; if USL $= \mu + 2\sigma$, LSL $= \mu - 2\sigma$, $C_P = 4/6 = 0.667$; and the process is potentially incapable.

Actual capability index C_{PK} is a measure of how well the actual process is meeting the product specification; it is defined by:

$$C_{PK} = \min\left\{\frac{\mu - LSL}{3\sigma}, \frac{USL - \mu}{3\sigma}\right\} \tag{1.15}$$

The values of C_{PK} have the same meaning as before: if C_{PK} is greater than 1 the process is meeting the specification, if C_{PK} is less than 1 the process is not meeting the specification. Thus if USL $= \mu + 4\sigma$, LSL $= \mu - 4\sigma$, $C_{PK} = \min\{1.33, 1.33\} = 1.33$, *i.e.* the same value as C_P above. However, C_{PK} has the advantage over C_P that it can detect a shift in the mean value μ of the characteristic away from the target value x_T. C_P cannot detect this shift and is therefore not suitable for measuring actual process capability. C_{PK} is calculated online from estimates of μ and σ.

Example 1.2

Due to process variability, a continuous performance characteristic x has a normal probability density function with mean μ and standard deviation σ. However, due to process drift, mean μ can be shifted from target value x_T by an amount Δ, *i.e.* $\mu = x_T + \Delta$ (Figure 1.6). Specification limits are USL $= x_T + 4\sigma$, LSL $= x_T - 4\sigma$. For $\Delta = 0$, 0.5σ, 1.0σ, 1.5σ, 2.0σ, 2.5σ, 3.0σ, calculate:

(a) C_P
(b) C_{PK}
(c) Probability of product being rejected.

Solutions

when $x = $ LSL $= x_T - 4\sigma$, $\mu = x_T + \Delta$

$$z = \frac{x - \mu}{\sigma} = \frac{x_T - 4\sigma - x_T - \Delta}{\sigma} = -4 - \frac{\Delta}{\sigma}$$

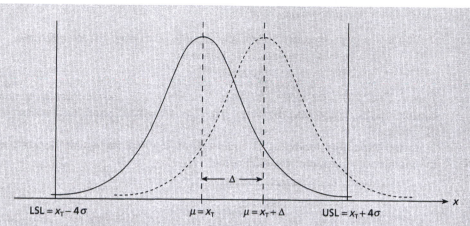

Figure 1.6   Shift in mean from target value

when $x = \text{USL} = x_T + 4\sigma$, $\mu = x_T + \Delta$

$$z = \frac{x - \mu}{\sigma} = \frac{x_T + 4\sigma - x_T - \Delta}{\sigma} = 4 - \frac{\Delta}{\sigma}$$

Answers to parts (a), (b) and (c) are given in Table 1.3, We note that as shift Δ increases, the probability of the product being rejected increases rapidly. This change in Δ is not detected by C_P which remains at 1.33. The change is detected by C_{PK} which decreases from 1.33 to 0.33 as the process is increasingly unable to meet the specification.

Table 1.3   Solution to Example 1.2

Δ	$C_P = \dfrac{\text{USL} - \text{LSL}}{6\sigma}$	$\dfrac{\mu - \text{LSL}}{3\sigma}$	$\dfrac{\text{USL} - \mu}{3\sigma}$	C_{PK}	$P_{x<\text{LSL}} = P_{-\infty,-4-\Delta/\sigma}$	$P_{x>\text{USL}} = P_{4-\Delta/\sigma,\infty}$	Probability of rejection
0	$\dfrac{8\sigma}{6\sigma} = 1.33$	$\dfrac{4\sigma}{3\sigma} = 1.33$	$\dfrac{4\sigma}{3\sigma} = 1.33$	1.33	$P_{-\infty,4} = 0.000\,03$	$P_{4,\infty} = 0.000\,03$	0.000 06
0.5σ	$\dfrac{8\sigma}{6\sigma} = 1.33$	$\dfrac{4.5\sigma}{3\sigma} = 1.50$	$\dfrac{3.5\sigma}{3\sigma} = 1.17$	1.17	$P_{-\infty,-4.5} \approx 0$	$P_{3.5,\infty} = 0.000\,23$	0.000 23
1.0σ	$\dfrac{8\sigma}{6\sigma} = 1.33$	$\dfrac{5\sigma}{3\sigma} = 1.67$	$\dfrac{3\sigma}{3\sigma} = 1.00$	1.0	$P_{-\infty,-5.0} \approx 0$	$P_{3.0,\infty} = 0.001\,35$	0.001 35
1.5σ	$\dfrac{8\sigma}{6\sigma} = 1.33$	$\dfrac{5.5\sigma}{3\sigma} = 1.83$	$\dfrac{2.5\sigma}{3\sigma} = 0.83$	0.83	$P_{-\infty,5.5} \approx 0$	$P_{2.5,\infty} = 0.006\,21$	0.006 21
2.0σ	$\dfrac{8\sigma}{6\sigma} = 1.33$	$\dfrac{6\sigma}{3\sigma} = 2.00$	$\dfrac{2\sigma}{3\sigma} = 0.67$	0.67	$P_{-\infty,6.0} \approx 0$	$P_{2.0,\infty} = 0.022\,75$	0.022 75
2.5σ	$\dfrac{8\sigma}{6\sigma} = 1.33$	$\dfrac{6.5\sigma}{3\sigma} = 2.17$	$\dfrac{1.5\sigma}{3\sigma} = 0.50$	0.50	$P_{-\infty,6.5} \approx 0$	$P_{1.5,\infty} = 0.066\,81$	0.066 81
3.0σ	$\dfrac{8\sigma}{6\sigma} = 1.33$	$\dfrac{7\sigma}{3\sigma} = 2.33$	$\dfrac{1\sigma}{3\sigma} = 0.33$	0.33	$P_{-\infty,7.0} \approx 0$	$P_{1.0,\infty} = 0.158\,66$	0.158 66

1.3.3 Quality control charts

Up to now quality control has been based on measuring a performance characteristic x for **all** individual items of a given product, *i.e.* for the entire **population** of items. Where the population is large this can be impractical, expensive and time consuming. It may be cheaper and quicker to make measurements on **samples** taken from the population; each sample will consist of a number of individuals. For many practical processes, the probability density function for values of x for the entire pooulation will not be normal or Gaussian. If, however, the average value $\bar{x}$ for each sample is calculated, then it can be shown that the probability density function for these averages is approximately normal or Gaussian. This is known as the Central Limit Theorem. The standard deviation of the variations in the averages $\bar{x}$ is less than the standard deviation σ of the variations in individual x values. However, because the averages follow an approximately normal distribution, it is possible to calculate **control limits** for average values $\bar{x}$, which correspond to the specification limits for individual values x discussed earlier.

Let us assume that the parent population of x values has a mean μ and a standard deviation σ. We can then make measurements on m sample sub-groups, each group consisting of n individuals, taken from the parent population at regular intervals. Typical values would be $m = 20$, $n = 5$. Suppose the ith value of x for the jth sub-group is x_{ij} where $i = 1, 2, \ldots, n, j = 1, 2, \ldots, m$. We first calculate **sub-group means** $\bar{x}_j$ using:

$$\bar{x}_j = \frac{1}{n} \sum_{i=1}^{n} x_{ij} \qquad (1.16)$$

For each sub-group j, we then find the maximum and minimum values of x for the group, these are $x_{\max j}$ and $x_{\min j}$: the **sub-group range** is then:

$$R_j = x_{\max j} - x_{\min j} \qquad (1.17)$$

Figures 1.7 (a) and (b) show values of $\bar{x}_j$, R_j plotted against sub-group number j; these are called **control charts**.

From the Central Limit Theorem, the probability density function for sub-group means $\bar{x}_j$ is approximately normal with mean $\bar{\bar{x}}$ and standard deviation $\sigma_{\bar{x}}$. Since this distribution is approximately normal, we can establish **control limits** on $\bar{x}_j$ values based on ± 3 standard deviations $\sigma_{\bar{x}}$, *i.e.*:

$$
\begin{aligned}
&\text{Upper Control Limit } \text{UCL}_{\bar{x}} = \mu + 3\sigma_{\bar{x}} \\
&\text{Lower Control Limit } \text{LCL}_{\bar{x}} = \mu - 3\sigma_{\bar{x}} \\
&\qquad\qquad \text{Centre line} = \mu
\end{aligned}
\qquad (1.18)
$$

These control limits correspond to the specification limits $\text{USL} = \mu + 3\sigma$, $\text{LSL} = \mu - 3\sigma$ on individual values x. Figure 1.8 shows the relationship between specification and control limits.

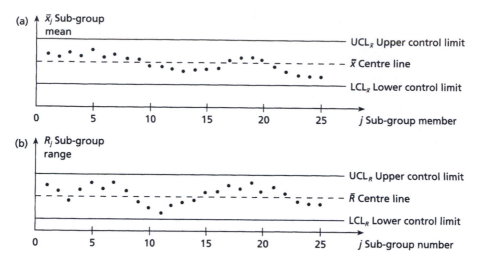

Figure 1.7   Control charts

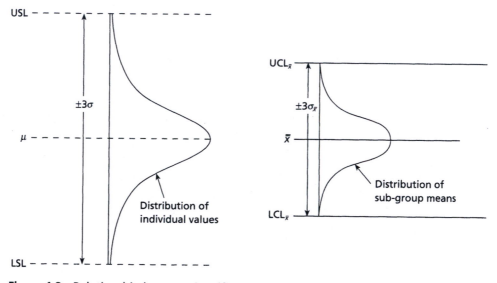

Figure 1.8   Relationship between Specification and Control limits

The population mean μ can be estimated from $\bar{\bar{x}}$, the mean of the sub-group means, *i.e.*

$$\mu \approx \bar{\bar{x}} = \frac{1}{m}\sum_{j=1}^{m}\bar{x}_j$$

(1.19)

The sub-group standard deviation $\sigma_{\bar{x}}$ can be estimated using:

$$3\sigma_{\bar{x}} = A_2\bar{R} \qquad (1.20)$$

Here A_2 is a constant which depends on the number of individuals n in each sub-group; for example, when $n = 5$, $A_2 = 0.58^2$. $\bar{R}$ is the mean of sub-group ranges and is given by:

$$\bar{R} = \frac{1}{m}\sum_{j=1}^{m} R_j \qquad (1.21)$$

The practical control limits on sub-group means $\bar{x}$ are therefore:

$$
\begin{aligned}
\text{UCL}_{\bar{x}} &= \bar{\bar{x}} + A_2\bar{R} \\
\text{LCL}_{\bar{x}} &= \bar{\bar{x}} - A_2\bar{R} \\
\text{Centre line} &= \bar{\bar{x}}
\end{aligned}
\qquad
\textbf{Practical control limits on } \bar{x} \textbf{ values}
\qquad (1.22)
$$

These control limits are also shown on the **x control chart** (Figure 1.7(a)). There are also corresponding practical control limits on sub-group R values; these are given by:

$$
\begin{aligned}
\text{UCL}_R &= D_4\bar{R} \\
\text{LCL}_R &= D_3\bar{R} \\
\text{Centre line} &= \bar{R}
\end{aligned}
\qquad
\textbf{Practical control limits on } R \textbf{ values}
\qquad (1.23)
$$

Here D_3, D_4 are constants which depend on the number of individuals n in each sub-group; for $n = 5$, $D_3 = 0.0$, $D_4 = 2.11$. These control limits are shown on the **R control chart** (Figure 1.7(b)). Thus for the process to be in statistical control, all $\bar{x}$ values must lie within the $\bar{x}$ control limits and all R values within the R control limits.

It is also possible to estimate the standard deviation σ for the entire population from the above mean range value $\bar{R}$, *i.e.*

$$\sigma_{\text{est}} = \frac{\bar{R}}{d_2} \qquad (1.24)$$

Here d_2 is a constant which again depends on the number of individuals n in each sub-group; for example, if $n = 5$, $d_2 = 2.326$, if $n = 10$, $d_2 = 3.078^2$. We can then use the estimates $\bar{\bar{x}}$ for μ and σ_{est} for σ in Equations 1.14 and 1.15 in order to estimate the capability indices C_P, C_{PK} for the process.

Example 1.3

A circular steel shaft is machined on a numerically controlled lathe; the target diameter for the shaft is 37.50 mm, the upper specification limit is 37.55 mm and the lower specification limit is 37.45 mm. Measurements x of shaft diameter are shown in Table 1.4; these are arranged in 25 sub-groups each containing 5 individuals. Only the two least significant figures are shown in the table.

(a) Estimate lower and upper control limits for means and ranges.
(b) Plot control charts for means and ranges.
(c) Estimate process capability index C_P.

Table 1.4 Measurements of shaft diameter[3]

Individual no. i \ Sub-group no. j	1	2	3	4	5	6	7	8	9	10	11	12	13
1	49	51	51	48	50	48	51	48	49	48	52	53	50
2	50	53	52	50	51	51	52	50	52	50	51	51	50
3	49	51	52	50	49	51	49	52	53	48	54	53	51
4	49	53	50	51	53	53	51	51	53	52	46	49	51
5	53	49	51	49	52	50	50	51	57	49	52	50	50
Means $\bar{x}_j$	50	51	51	50	51	51	51	50	52	49	51	51	50
Ranges R_j	04	04	02	03	04	05	03	04	04	04	08	04	01

Individual no. i \ Sub-group no. j	14	15	16	17	18	19	20	21	22	23	24	25
1	49	54	51	51	51	49	48	46	48	52	49	49
2	50	52	52	50	50	49	48	47	52	49	48	50
3	53	52	51	53	48	53	49	49	49	51	49	50
4	51	51	52	50	50	49	48	52	49	49	48	48
5	48	53	49	52	48	50	49	52	51	49	50	49
Means $\bar{x}_j$	50	52	51	51	49	50	48	49	50	50	49	49
Ranges R_j	05	03	03	03	03	04	01	06	04	03	02	02

Solutions

(a) Table 1.4 shows means $\bar{x}_j$ and ranges R_j for each sub-group j

Mean of sub-group means $\bar{\bar{x}} = \dfrac{1}{25}\displaystyle\sum_{j=1}^{25} \bar{x}_j = 37.502$ mm

Mean of sub-group ranges $\bar{R} = \dfrac{1}{25}\displaystyle\sum_{j=1}^{25} R_j = 0.036$ mm

For $n = 5$, $A_2 = 0.58$, $D_3 = 0.0$, $D_4 = 2.11$

Upper Control Limit on $\bar{x}$, $\text{UCL}_{\bar{x}} = \bar{\bar{x}} + A_2\bar{R} = 37.502 + 0.58(0.036)$
$$= \textbf{37.523 mm}$$

Lower Control Limit on $\bar{x}$, $\text{LCL}_{\bar{x}} = \bar{\bar{x}} - A_2\bar{R} = 37.502 - 0.58(0.036)$
$$= \textbf{37.481 mm}$$
$$\text{Centre line} = \bar{\bar{x}} = \textbf{37.502 mm}$$

Upper Control Limit on R, $\text{UCL}_R = D_4\bar{R} = 2.11(0.036) = \textbf{0.076 mm}$
Lower Control Limit on R, $\text{LCL}_R = D_3\bar{R} = 0.0(0.036) = \textbf{0.0 mm}$
$$\text{Centre line} = \bar{R} = \textbf{0.036 mm}$$

(b)

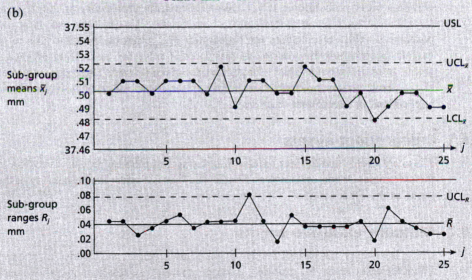

Figure 1.9 Control Chart

(c) $\sigma_{\text{est}} = \dfrac{\bar{R}}{d_2}$, $d_2 = 2.326$ if $n = 5$

$$= \dfrac{0.036}{2.326} = 0.0155 \text{ mm}$$

$$C_p = \frac{USL - LSL}{6\sigma_{est}} = \frac{(37.55 - 37.45)}{6 \times 0.0155} = \frac{0.10}{0.093}$$

$$C_P = 1.08$$

This means that the process is just capable of meeting the specification; this is confirmed by inspection of the control charts.

1.4 The combination of probabilities

In Section 1.2 we saw that the quality of a product can be specified by a set of continuous performance characteristics $x_1, x_2, \ldots$, *i.e.* a vector $\{x\}$. The specification for the product can then be defined in terms of a target vector $\{x_T\}$ a lower specification limit vector $\{LSL\}$ and an upper specification limit vector $\{USL\}$. In Section 1.3.1 we saw how to calculate the probability P that a **single** performance characteristic x will lie outside the specification limits LSL, USL. We must calculate this probability for each perfomance characteristic, *i.e.* for each element of $\{x\}$. The result of these calculations will be a set of probabilities $\{P\}$, each one giving the probability of failure to meet the specification for a given performance characteristic. We can then use these individual probability values to calculate the probability $P_{overall}$ that the product fails to meet the overall specification. In order to do this a set of rules for calculating an overall probability from several individual probabilities are given below[4]. These rules will be applied to the above quality problem in Example 1.4 and to a range of reliability and quality applications in subsequent chapters.

1.4.1 Independent events

These are events that are not related in any way and may be defined as events in which the occurrence or non-occurrence of one event does not affect the probability of the occurrence of the other event.

1.4.2 Mutually exclusive events

These are events that cannot happen at the same time: the occurrence of one event prohibits the other event; for example, if a robot moves forwards it cannot move backwards.

1.4.3 Complementary events

These are events such that if event A does not occur, event B must occur and vice versa; for example, if a tossed coin is not heads it must be tails. The sum of probabilities for complementary events must therefore be 1, *i.e.*

$$\boxed{P_A + P_B = 1}$$ **Sum of probabilities for complementary events** (1.25)

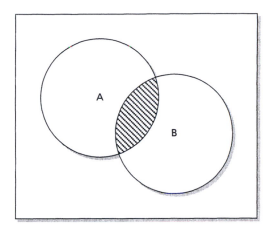

Figure 1.10 Venn diagram for simultaneous events

1.4.4 Simultaneous events – product rule

Suppose that A and B are independent events which are non-mutually exclusive and that P_A is the probability of A occurring and P_B the probability of B occurring. The probabilities P_A and P_B are represented by the areas of the circles A and B in the Venn diagram (Figure 1.10). The probability of **both** events A **and** B occurring is equal to the **product** of the individual probabilities, *i.e.*

$$P_{\text{A AND B}} = P_A \times P_B \qquad \text{product rule for simultaneous events} \qquad (1.26)$$

and is represented by the shaded overlap area in Figure 1.10. For mutually exclusive events, $P_A P_B = 0$ and the overlap area is zero. In general, if there are n independent non-mutually exclusive events with probabilities $P_1, P_2, \ldots, P_i, \ldots, P_n$, then the probability of event 1 **and** event 2 . . . **and** event i . . . **and** event n occurring is given by:

$$P_{1 \text{ AND } 2 \ldots \text{AND } i \ldots \text{AND } n} = P_1 \times P_2 \times \ldots \times P_i \times \ldots \times P_n \qquad \text{product rule} \qquad (1.27)$$

1.4.5 Occurrence of at least one of several events – addition rule

Given the probabilities P_A and P_B of the individual independent events A and B occurring, we also need to know the probability of either event A **or** event B **or both** events occurring. This is represented by the total area of the overlapping circles in Figure 1.10 and is given by:

$$P_{\text{A OR B OR BOTH}} = P_A + P_B - P_A P_B \qquad \text{probability of either event or both} \qquad (1.28)$$

i.e. by the sum of the areas of the individual circles less the area of overlap. If the events are also mutually exclusive, then the probability $P_A P_B$ of **both** events occurring is zero; the probability of A **or** B occurring is then given by the **sum** of the individual probabilities:

$$P_{A\,\mathrm{OR}\,B} = P_A + P_B$$ **probability of either event – addition rule**   (1.29)

In general, if there are n independent, mutually exclusive events with probabilities P_1, $P_2, \ldots, P_i, \ldots, P_n$, then the probability of event 1 **or** event 2 . . . **or** event i . . . **or** event n occurring is given by:

$$P_{1\,\mathrm{OR}\,2\,\ldots\,\mathrm{OR}\,i\,\ldots\,\mathrm{OR}\,n} = P_1 + P_2 + \ldots + P_i + \ldots + P_n$$ **addition rule**   (1.30)

The above addition rule may also be approximately true for n independent, non-mutually exclusive events where the probabilities of the events occurring simultaneously is small. For example, if we have $P_A = P_B = 0.1$ and $P_A P_B = 0.01$ then from Equation 1.28 we have

$$P_A + P_B - P_A P_B = 0.1 + 0.1 - 0.01 = 0.19$$

and from Equation 1.29.

$$P_A + P_B = 0.20$$

i.e. the error introduced by using the addition rule is only 5%. This is called the **rare events approximation**.

1.4.6 The binomial distribution

The binomial distribution is a **discrete** probability distribution which applies in situations where n independent trials are made and there are only two possible outcomes to each trial. If, for example, a coin is tossed ten times, the binomial distribution can be used to calculate the probability of getting heads ten times or tails three times. The binomial distribution is therefore very useful in reliability engineering where there are only two possible outcomes, survival with probability R and failure with probability F. The binomial distribution is obtained from the polynomial expansion of $(R + F)^n$, *i.e.*

$$(R + F)^n = R^n + nR^{n-1}F + \frac{n(n-1)}{2!}R^{n-2}F^2 + \ldots$$

$$+ \frac{n(n-1)(n-2)\ldots(n-j+1)}{j!}R^{n-j}F^j + \ldots + nRF^{n-1} + F^n$$

$$= \sum_{j=0}^{j=n} {}^nC_j R^{n-j}F^j$$ (1.31)

where ${}^nC_j = n!/[(n-j)!\,j!]$. Since the events are complementary, $R + F = 1$.

The first term R^n represents the probability of n survivals in n trials, the second term $nR^{n-1}F$ the probability of $n-1$ survivals in n trials, and the jth term $^nC_jR^{n-j}F^j$ the probability of $n-j$ survivals in n trials. Examples of the use of the binomial distributions are given in Section 3.5, Example 3.2 and in Section 7.5.2.

Example 1.4

Table 1.5 lists a set $\{x\}$ of performance characteristics for a family car each of which can be represented by a normal probability density function. The table gives corresponding sets of target values $\{x_T\}$, lower specification limits $\{LSL\}$, upper specification limits $\{USL\}$, mean values $\{\mu\}$ and standard deviations $\{\sigma\}$.

(a) Use the $P_{u,\infty}$ values given in Table 1.2 and the properties of the normal distribution, to calculate the set of probabilities $\{P\}$, that the car will fail to meet the specification for each performance characteristic.

(b) Assuming that the probabilities of simultaneous failures are negligible, estimate the overall probability of failure to meet the specification.

Solutions

(a) From Section 1.3.1, the probability that the characteristic fails to meet the specification is:

$$P = P_{x>USL} + P_{x<LSL} = P_{USL,\infty} + P_{-\infty,LSL}$$

(i) P_1

When $x = USL = 42.5$, $u = \dfrac{x - \mu}{\sigma} = \dfrac{42.5 - 40}{1.0} = 2.5$

$$P_{USL,\infty} = P_{2.5,\infty} = 0.0062$$

When $x = LSL = 37.5$, $u = \dfrac{37.5 - 40}{1.0} = -2.5$

$$P_{-\infty,LSL} = P_{-\infty,-2.5} = P_{2.5,\infty} = 0.0062$$
$$P_1 = 0.0062 + 0.0062 = 0.0124$$

Table 1.5 Performance characteristics and specification limits

Performance Characteristics	Target x	Lower Spec. Limit LSL	Upper Spec. Limit USL	Mean	Standard Deviation
Urban m.p.g.	40	37.5	42.5	40	1.0
Maximum m.p.h.	100	94	106	100	3.0
Time to 60 m.p.h., seconds	12	9.5	14.5	12.5	1.0
Braking dist. at 60 m.p.h., ft	180.0	178.75	181.25	179.75	0.5

(ii) P_2

When $x =$ USL $= 106$, $u = \dfrac{106 - 100}{3.0} = 2.0$

$P_{USL,\infty} = P_{2.0,\infty} = 0.0228$

When $x =$ LSL $= 94$, $u = \dfrac{94 - 100}{3.0} = -2.0$

$P_{-\infty, LSL} = P_{-\infty, -2.0} = P_{2.0,\infty} = 0.0228$

$P_2 = 0.0228 + 0.0228 = \mathbf{0.0456}$

(iii) P_3

When $x =$ USL $= 14.5$, $u = \dfrac{14.5 - 12.5}{1.0} = 2.0$

$P_{USL,\infty} = P_{2.0,\infty} = 0.0228$

When $x =$ LSL $= 9.5$, $u = \dfrac{9.5 - 12.5}{1.0} = -3.0$

$P_{-\infty, LSL} = P_{-\infty, -3.0} = P_{3.0,\infty} = 0.0014$

$P_3 = 0.0228 + 0.0014 = \mathbf{0.0242}$

(iv) P_4

When $x =$ USL $= 181.25$, $u = \dfrac{181.25 - 179.75}{0.5} = 3.0$

$P_{USL,\infty} = P_{3.0,\infty} = 0.0014$

When $x =$ LSL $= 178.75$, $u = \dfrac{178.75 - 179.75}{0.5} = -2.0$

$P_{-\infty, LSL} = P_{-\infty, -2.0} = P_{2.0,\infty} = 0.0228$

$P_4 = 0.0228 + 0.0014 = \mathbf{0.0242}$

(b) The car will fail to meet the specification if **any individual** performance characteristic fails to meet the specification for that characteristic. The **overall** probability $P_{OVERALL}$ of failure to meet the specification is therefore given by:

$P_{OVERALL} =$ Probability P_1 that x_1 fails specification

OR

Probability P_2 that x_2 fails specification

OR

Probability P_3 that x_3 fails specification

etc.

Assuming that the probabilities $P_1 \times P_2 \times P_3$ etc. of simultaneous failures are negligible, then $P_{OVERALL}$ is the **sum** of the individual failure probabilities (Equation 1.29), *i.e.*:

$P_{OVERALL} = P_1 + P_2 + P_3 + P_4$
$= 0.0124 + 0.0456 + 0.0242 + 0.0242 = \mathbf{0.1064}$

Summary

This chapter explained and defined the concept of quality. It then discussed statistical techniques for quantifying and controlling quality.

References

1. BS 5750 (1987) *Quality Systems, part 0: Principal Concepts and Applications*, British Standards Institution, London.
2. Kolarik, W. J. (1995) *Creating Quality: Concepts, Systems, Strategies and Tools*, McGraw-Hill, New York, p. 879.
3. Mamzic, C. (1995) Guidelines for the application of Statistical Process Control in the process industries, *Measurement and Control*, **28**, April, pp. 69–73.
4. Thomson, J. R. (1987) *Engineering Safety Assessment*, Longman Scientific and Technical, Harlow, pp. 10–13.

Self-assessment questions

1.1 A quality test on a component yielded the following 35 values of a continuous performance characteristic:

208.6; 208.3; 208.7; 208.5; 208.8; 207.6; 208.9; 209.1; 208.2; 208.4; 208.1; 209.2; 209.6; 208.6; 208.5; 207.4; 210.2; 209.2; 208.7; 208.4; 207.7; 208.9; 208.7; 208.0; 209.0; 208.1; 209.3; 208.2; 208.6; 209.4; 207.6; 208.1; 208.8; 209.2; 209.7

(a) Using equal intervals of 0.5, plot:
  (i)   the discrete probability distribution
  (ii) the discrete probability density distribution for the data.
(b) Calculate the mean and standard deviation for the data.
(c) Sketch a normal probability density function with the mean and standard deviation calculated in (b) on the distribution drawn in (a) (ii).

1.2 A normal probability density function $p(x)$ has a mean value $\mu = 25$ and a standard deviation $\sigma = 5$. Calculate the probability of x:

(a) being greater than 40
(b) being less than 15
(c) lying between 15 and 40
(d) lying between 20 and 30
(e) lying outside 20 and 30.

1.3 As a result of process variations, the resistance of a precision resistor follows a normal distribution with mean 50.0 Ω and standard deviation 0.1 Ω. The target value of resistance is 50.0 Ω with a lower specification limit of 49.65 Ω and an upper specification limit of 50.35 Ω.

(a) Calculate:
  (i)   Potential capability index C_P
  (ii)   Actual capability index C_{PK}
  (iii) Probability of the product being rejected.
(b) How are the above values changed if the mean value chages to 49.9 Ω due to drift in process conditions?

1.4 A component is machined on a numerically controlled milling machine. The target value for a particular dimension is 55.00 mm, the upper specification limit is 55.25 mm and the lower specification limit is 54.75 mm. Measurements of the dimension in mm are shown in Table Q1.4; these are arranged in 10 sub-groups each containing five individuals. Using information given in Section 1.3.1:

(a) estimate lower and upper control limits for means and ranges;
(b) plot control charts for means and ranges;
(c) estimate potential capability index C_P;
(d) comment on the above results.

Table Q1.4

Individual no. \ Sub-group no.	1	2	3	4	5	6	7	8	9	10
1	55.21	55.13	55.11	54.93	54.88	55.22	55.11	55.02	55.12	55.02
2	54.99	55.03	55.01	55.12	55.14	55.15	55.07	55.06	55.03	54.97
3	55.10	54.98	54.99	55.04	55.05	54.97	54.99	54.97	54.95	55.06
4	55.02	54.96	54.97	54.98	55.11	54.95	54.93	54.99	54.98	54.99
5	54.95	55.04	54.98	54.91	54.97	55.02	55.00	54.01	54.93	55.01

1.5 The probabilities of five non-mutually exclusive events occurring are:
  0.01, 0.02, 0.04, 0.05, 0.08

(a) What is the maximum probability of any two events occurring simultaneously?
(b) What is the maximum probability of any three events occurring simultaneously?
(c) Estimate the probability of any one of the five events occurring.

1.6 The probability of a missile hitting a target is 0.95.

(a) If two salvoes are fired, calculate the probability of:
  (i)   two hits
  (ii)   two misses
  (iii) at least one hit.
(b) If three salvoes are fired, use the expansion of $(R + F)^3$ to calculate the probability of:

    (i)   three hits
    (ii)   at least two hits
    (iii)   at least one hit
    (iv)   three misses.

1.7 A coin is tossed 10 times. Assuming that for each toss there is an equal probability of getting heads or tails, calculate the probability of:
(a) getting heads 10 times
(b) getting tails 10 times
(c) getting heads six times
(d) getting tails three times.

2

Principles of reliability

2.1 Introduction

The previous chapter has explained the meaning of **quality** and shown how it can be quantified. The word **reliability** is also commonly used in everyday life. When a product such as a car or washing machine breaks down, the user is forcibly made aware of the limited reliability of the product. The aim of this chapter is to define the concept of reliability and to explain how it is quantified.

2.2 The meaning of reliability[1]

Suppose that a newly manufactured individual item of a given product is tested, either before despatch from the manufacturer, or on receipt by the user, or both, and the performance characteristics $\{x\}$ are found to satisfy the specification $\{x_T\}$, $\{\Delta\}$. The product is then placed in service. If, as time goes on, the product continues to meet the specification, then it is considered to have survived. **The reliability R of the product can therefore be defined as the probability that the product continues to meet the specification**, over a given time period, subject to given environmental conditions. If, however, as time goes on the product fails to meet the specification, then it is considered to have failed. **The unreliability F of the product can be defined as the probability that the product fails to meet the specification**, over a given time period, subject to given environmental conditions. Failure can occur due to many factors; examples are wear, mechanical fracture and chemical corrosion.

Both reliability and unreliability vary with time. Reliability $R(t)$ decreases with time: an item that has just been tested and shown to meet specification has a reliability of 1 when first placed in service; one year later this may have decreased to 0.5. Unreliability $F(t)$ increases with time; an item that has just been tested and shown to meet specification has an unreliability of 0 when first placed in service, increasing to say 0.5 after one year. Since, at any time t, the product has either survived or failed, the sum of reliability and unreliability must be 1, *i.e.* the events are **complementary** and:

$$R(t) + F(t) = 1 \qquad (2.1)$$

We can now discuss the relationship between quality and reliability. The reliability of a product is its ability to retain its quality as time progresses. Thus a product can only have high quality if it also has high reliability; high initial quality is of little use if it is soon lost. The opposite is, however, not true; a product with high reliability does not necessarily have high quality, but may be merely retaining low quality over a long period of time.

2.3 Practical reliability definitions

$R(t)$ and $F(t)$ are dependent on time; it is useful to have measures of reliability which are independent of time. We will consider two cases; in the first the items are non-repairable and in the second the items are repairable.

2.3.1 Non-repairable items

Suppose that N individual items of a given non-repairable product are placed in service and the times at which failures occur are recorded during a test interval T. We further assume that all the N items fail during T and that the ith failure occurs at time T_i, that is, T_i is the survival time or **up time** for the ith failure. The total up time for N failures is therefore $\sum_{i=1}^{i=N} T_i$ and the **mean time to failure** is given by:

$$\textbf{mean time to fail} = \frac{\text{Total up time}}{\text{Number of failures}}$$

i.e.

$$\text{MTTF} = \frac{1}{N} \sum_{i=1}^{i=N} T_i \qquad (2.2)$$

The mean failure rate $\bar{\lambda}$ is correspondingly given by:

$$\textbf{mean failure rate} = \frac{\text{Number of failures}}{\text{Total up time}}$$

i.e.

$$\bar{\lambda} = \frac{N}{\sum_{i=1}^{i=N} T_i} \tag{2.3}$$

i.e. mean failure rate is the reciprocal of MTTF.

There are N survivors at time $t = 0$, $N - i$ at time $t = T_i$, decreasing to zero at time $t = T$; Figure 2.1(a) shows how the probability of survival, *i.e.* reliability, $R_i = (N - i)/N$ decreases from $R_i = 1$ at $t = 0$, to $R_i = 0$ at $t = T$. The ith rectangle has height $1/N$ and length T_i and area T_i/N. Therefore from Equation 2.2 we have:

MTTF = Total area under the graph

In the limit that $N \to \infty$, the discrete reliability function R_i becomes the continuous function $R(t)$. The area under $R(t)$ is $\int_0^T R(t)dt$ so that we have in general:

$$\text{MTTF} = \int_0^\infty R(t)dt \tag{2.4}$$

The upper limit of $t = \infty$ corresponds to N being infinite.

2.3.2 Repairable items

Figure 2.1(b) shows the failure pattern for N items of a repairable product observed over a test interval T. The **down time** T_{Dj} associated with the jth failure is the total time that elapses between the occurrence of the failure and the repaired item being put back into normal operation. The total down time for N_F failures is therefore $\sum_{j=1}^{j=N_F} T_{Dj}$ and the **mean down time** is given by:

$$\textbf{mean down time} = \frac{\text{Total down time}}{\text{Number of failures}}$$

i.e.

$$\text{MDT} = \frac{1}{N_F} \sum_{j=1}^{j=N_F} T_{Dj} \tag{2.5}$$

The total **up time** can be found by subtracting the total down time from NT, *i.e.*

$$\textbf{total up time} = NT - \sum_{j=1}^{j=N_F} T_{Dj}$$

$$= NT - N_F \text{MDT}$$

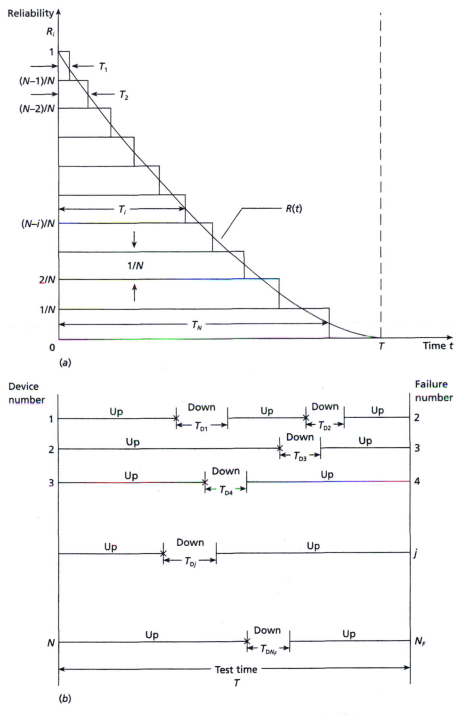

Figure 2.1 Failure patterns: (a) non-repairable items; (b) repairable items

The mean up time or the **mean time between failures (MTBF)** is therefore given by:

$$\text{mean time between failures} = \frac{\text{Total up time}}{\text{Number of failures}}$$

i.e.

$$\boxed{\text{MTBF} = \frac{NT - N_F\text{MDT}}{N_F}}$$

(2.6)

The **mean failure rate** $\bar{\lambda}$ is correspondingly given by:

$$\text{mean failure rate} = \frac{\text{Number of failures}}{\text{Total up time}}$$

i.e.

$$\boxed{\bar{\lambda} = \frac{N_F}{NT - N_F\text{MDT}}}$$

(2.7)

i.e. again mean failure rate is the reciprocal of **MTBF**.

2.4 The meaning of availability

When a repairable product is **up**, *i.e.* working satisfactorily, it is available for use. When the product is **down**, *i.e.* being repaired, it is unavailable for use. It is important to have an average measure of the degree to which the product is either available or unavailable.

The **availability** of the product is the fraction of the total test interval that it is performing within specification, *i.e.* up; thus we have:

$$\begin{aligned}
\text{availability} &= \frac{\text{Total up time}}{\text{Test interval}} \\
&= \frac{\text{Total up time}}{\text{Total up time} + \text{Total down time}} \\
&= \frac{N_F \times \text{MTBF}}{N_F \times \text{MTBF} + N_F \times \text{MDT}}
\end{aligned}$$

i.e.

$$\boxed{A = \frac{\text{MTBF}}{\text{MTBF} + \text{MDT}}}$$

(2.8)

Unavailability U is similarly defined as the fraction of the total test interval that it is not performing to specification, *i.e.* failed or down, thus we have:

$$\textbf{unavailability} = \frac{\text{Total down time}}{\text{Test interval}}$$

giving:

$$U = \frac{\text{MDT}}{\text{MTBF} + \text{MDT}} \tag{2.9}$$

It follows from Equations 2.8 and 2.9 that:

$$A + U = 1 \tag{2.10}$$

We see from Equations 2.8 and 2.9 that A and U depend on MTBF, *i.e.* availability depends on **reliability**. Availability can therefore be increased by increasing MTBF, *i.e.* reducing mean failure rate. We see also that A and U depend on Mean Down Time, MDT, availability can be increased by reducing MDT. Thus availability also depends on **maintainability**, *i.e.* how quickly the product can be repaired and put back into service. Maintenance is discussed in detail in Chapter 6.

2.5 Instantaneous failure rate and its relation to reliability

We assume to begin with that n items of a product survive up to time $t = \xi$ and that Δn items fail during the small time interval $\Delta \xi$ between ξ and $\xi + \Delta \xi$. The probability of failure during interval $\Delta \xi$ (given survival to time ξ) is therefore equal to $\Delta n/n$. Assuming no repair during $\Delta \xi$ the corresponding **instantaneous failure rate or hazard rate** at time ξ is, from Equation 2.3, given by:

$$\lambda(\xi) = \frac{\Delta n}{n \Delta \xi} = \frac{\text{Failure probability}}{\Delta \xi} \tag{2.11}$$

Comparing Equation 2.11 with Equations 1.6 and 1.8, we note that $\lambda(\xi)$ is effectively a **probability density function**, but is a function of time ξ rather than performance characteristic x.

The unconditional probability ΔF that an item fails during the interval $\Delta \xi$ is:

ΔF = Probability that item survives up to time ξ

and

Probability that item fails between ξ and $\xi + \Delta \xi$ (given survival to ξ)

The first probability is given by $R(\xi)$ and from Equation 2.11, the second probability is $\lambda(\xi) \Delta \xi$. From the **product** rule of Section 1.4.4 we have:

$$\Delta F = R(\xi)\lambda(\xi)\Delta \xi$$

i.e.

$$\frac{\Delta F}{\Delta \xi} = R(\xi)\lambda(\xi)$$

Thus in the limit that $\Delta \xi \to 0$, we have:

$$\frac{dF}{d\xi} = R(\xi)\lambda(\xi) \tag{2.12}$$

also since $F(\xi) = 1 - R(\xi)$, $dF/d\xi = -(dR/d\xi)$, giving:

$$-\frac{dR}{d\xi} = R(\xi)\lambda(\xi)$$

i.e.

$$\int_{R(0)}^{R(t)} \frac{dR}{R} = -\int_0^t \lambda(\xi)d\xi \tag{2.13}$$

In Equation 2.13, the left hand integral is with respect to R and the right hand integral with respect to t. Since at $t = 0$, $R(0) = 1$, we have:

$$[\log_e R]_1^{R(t)} = -\int_0^t \lambda(\xi)d\xi$$

$$\log_e R(t) = -\int_0^t \lambda(\xi)d\xi$$

i.e.

$$\boxed{R(t) = \exp\left[-\int_0^t \lambda(\xi)d\xi\right]}$$

relation between reliability and instantaneous failure rate $\tag{2.14}$

2.6 | Typical forms of hazard rate function

In the previous section instantaneous failure rate or hazard rate $\lambda(t)$ was defined. Figure 2.2 shows the most general form of $\lambda(t)$ throughout the lifetime of a product. This is the so-called bathtub curve and consists of three distinct phases: early failure, useful life and wear-out failure. The **early failure region** is characterized by $\lambda(t)$ decreasing with time. When items are new, especially if the product is a new design, early failures can occur due to design faults, poor quality components, manufacturing faults, installation errors, and operator and maintenance errors; the latter may be due to unfamiliarity with the product. The hazard rate falls as design faults are rectified, weak components are removed and the user becomes familiar with installing, operating and maintaining the product. The **useful life region** is characterized by a low, constant failure rate. Here all weak components have been removed: design, manufacture, installation, operating and maintenance errors rectified so that failure is due to a variety of unpredictable causes. The **wear-out region** is characterized by $\lambda(t)$ increasing with time as individual items approach the end of the design life for the product; long-life components which make up the product are now wearing out.

While the bathtub curve represents the most general form of the time variation in instantaneous failure rate, the failure of many products and components can be satisfactorily represented by simpler forms of $\lambda(t)$. The most commonly used are the **Constant Failure Rate** and **Weibull Failure Rate** models.

2.6.1 Constant failure rate and the exponential distribution

A study of the failure rate of a large range of aerospace industry products and components showed that six different forms of $\lambda(t)$ could be identified; 68% of all the products

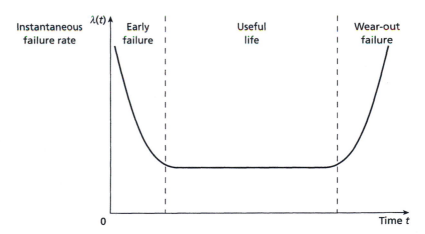

Figure 2.2 Typical variation in instantaneous failure rate (hazard rate) during the lifetime of product – 'bathtub curve'

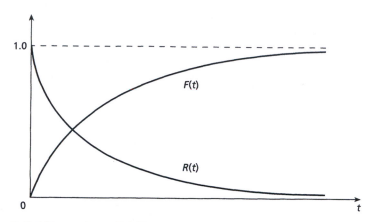

Figure 2.3   Reliability and unreliability with constant failure model

could be represented by a $\lambda(t)$ characterized by a short early failure region, an extended constant failure region and no wear-out region. Many electronic devices fall into this category. Also, the use of effective quality control techniques such as 'burn-in' (Chapter 5) means that the early failure region can either be reduced or eliminated. Thus a constant failure rate model will be adequate for a large range of products and components; here we have:

$$\lambda(t) = \lambda(\xi) = \lambda = \text{constant} \tag{2.15}$$

so that:

$$R(t) = \exp\left[-\lambda \int_0^t \xi\right] = \exp(-\lambda t)$$

and: $\qquad\qquad\qquad\qquad\qquad\qquad\qquad\qquad\qquad\qquad\qquad\qquad\qquad$ (2.16)

$$F(t) = 1 - \exp(-\lambda t)$$

Thus a constant failure or hazard rate gives rise to an **exponential** reliability time variation or distribution (Figure 2.3).

From Equations 2.14 and 2.16 the mean time to failure in the constant rate case is given by:

$$\text{MTTF} = \int_0^\infty R(t)\,\mathrm{d}t = \int_0^\infty \exp(-\lambda t)\,\mathrm{d}t = \left[-\frac{1}{\lambda}\exp(-\lambda t)\right]_0^\infty$$

$$= -\frac{1}{\lambda}[0 - 1] = \frac{1}{\lambda} \tag{2.17}$$

Thus MTF is the reciprocal of failure rate in the constant failure rate case.

2.6.2 Weibull failure rate and distribution

The Weibull instantaneous failure rate or hazard rate function is given by:

$$\lambda(t) = \boxed{\frac{\beta}{\eta}\left(\frac{t - t_0}{\eta}\right)^{\beta-1}} \quad \beta > 0 \tag{2.18}$$

The function is therefore defined by three parameters t_0, η, β; t_0 determines the position of the origin (reliability $R(t_0) = 1$ at $t = t_0$), η is the scale parameter and β the shape parameter. Using Equation 2.14 the corresponding reliability time variation or distribution is given by:

$$R(t) = \exp\left\{-\int_{t_0}^{t} \lambda(\xi)\mathrm{d}\xi\right\} = \exp\left\{-\frac{\beta}{\eta^\beta}\int_{t_0}^{t}(\xi - t_0)^{\beta-1}\mathrm{d}\xi\right\}$$

$$= \exp\left\{-\frac{\beta}{\eta^\beta}\left[\frac{1}{\beta}(\xi - t_0)^\beta\right]_{t_0}^{t}\right\} = \exp\{-1/\eta^\beta\,[(t - t_0)^\beta - 0]\}$$

i.e.

$$R(t) = \boxed{\exp\left\{-\left(\frac{t - t_0}{\eta}\right)^\beta\right\}} \tag{2.19}$$

If we make the substitution $\tau = (t - t_0)/\eta$, where τ is normalized time, Equations 2.18 and 2.19 simplify to the normalized pair:

$$\eta\lambda(\tau) = \beta\tau^{\beta-1}, \quad R(\tau) = \exp(-\tau^\beta) \tag{2.20}$$

Figures 2.4(a) and (b) show how the normalized failure rate function $\eta\lambda(\tau)$ and reliability function $R(\tau)$ vary with normalized time τ, for different values of shape parameter β. We see that for $\beta < 1$, $\eta\lambda(\tau)$ decreases as τ increases corresponding to early failure. For $\beta = 1$, $\eta\lambda(\tau) = 1$ for all τ, corresponding to constant failure rate; for $\beta > 1$, $\eta\lambda(\tau)$ increases as τ increases corresponding to wear-out failure. Thus the Weibull function can be used to respresent all three regions shown in Figure 2.2.

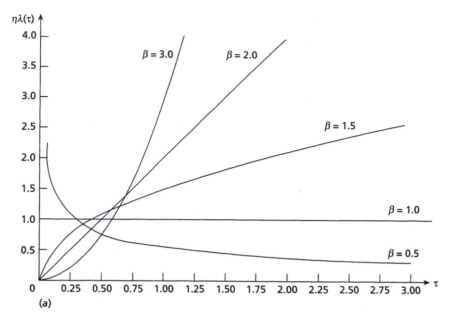

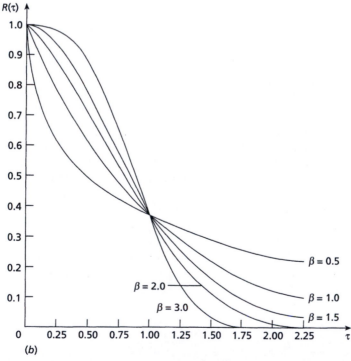

Figure 2.4   Weibull failure rate model: (a) normalized failure rate $\eta\lambda(\tau) = \beta\tau^{\beta-1}$; (b) normalized reliability function $R(\tau) = \exp(-\tau^\beta)$

Example 2.1

The reliability of 1000 non-repairable components is to be measured over a total time interval of 50 000 hours. At the beginning of the test interval all components are working to specification. Table 2.1 shows how the cumulative total number of failures increases with time. Using this data:

(a) Find the time variation of unreliability $F(t)$ and reliability $R(t)$.
(b) Decide whether the data follows a Weibull distribution.
(c) Estimate the values of t_0, η and β.
(d) Find the probability of survival to $t = 10\,000$ hours.
(e) Find an equation for instantaneous hazard rate $\lambda(t)$; is $\lambda(t)$ decreasing, increasing or constant with time?
(f) Estimate mean time to failure, MTTF.

Table 2.1 Data for Example 2.1

Time t $\times 10^3$ hours	Cumulative total number of failures n	Unreliability $F(t) = n/1000$	Reliability $R(t) = 1 - F(t)$
0	0	0.000	1.000
0.75	22	0.022	0.978
0.80	30	0.030	0.970
0.90	36	0.036	0.964
1.4	42	0.042	0.958
1.5	58	0.058	0.942
2.0	74	0.074	0.926
2.3	105	0.105	0.895
3	140	0.140	0.860
5	200	0.200	0.800
6	290	0.290	0.710
8	350	0.350	0.650
11	540	0.540	0.460
15	570	0.570	0.430
19	770	0.770	0.230
37	920	0.920	0.080

Solutions

(a) See Table 2.1.
(b) Weibull reliability

$$R(t) = \exp\left\{-\left(\frac{t - t_0}{\eta}\right)^{\beta}\right\} = 1 - F(t)$$

i.e.

$$\frac{1}{1 - F(t)} = \exp\left\{\left(\frac{t - t_0}{\eta}\right)^{\beta}\right\}$$

i.e.

$$\log_e\left[\frac{1}{1 - F(t)}\right] = \left(\frac{t - t_0}{\eta}\right)^{\beta}$$

and

$$\log_e\left\{\log_e\left[\frac{1}{1 - F(t)}\right]\right\} = \beta \log_e\left(\frac{t - t_0}{\eta}\right)$$

i.e.

$$\log_e \log_e\left[\frac{1}{1 - F(t)}\right] = \beta \log_e(t - t_0) - \beta \log_e \eta$$

Thus for a Weibull distribution, a graph of $\log_e \log_e\left[\dfrac{1}{1 - F(t)}\right]$ versus $\beta \log_e(t - t_0)$ will be a straight line of the form:

$$y = mx + c$$

where gradient $m = \beta$

and intercept $\quad c = \beta \log_e \eta$

Weibull probability graph paper is constructed by having a vertical (y) axis proportional to the log log of the reciprocal $1/F$ of unreliability F and a horizontal (x) axis proportional to the log of time t. Figure 2.5 shows the data of Table 2.1 plotted on Weibull graph paper; a straight line graph is obtained confirming that the data follows a Weibull distribution.

(c)   From Table 2.1, when $t = 0$, $R(0) = 1$, *i.e.*

$$1 = \exp\{-(t_0/\eta)^{\beta}\} \quad \textit{i.e. } \boldsymbol{t_0 = 0}$$

i.e.

$$\boldsymbol{R(t) = \exp\{-(t/\eta)^{\beta}\}}$$

when $t = \eta$, *i.e.* $t/\eta = 1$, $R(t) = e^{-1} = 0.368$ and $F(t) = 1 - e^{-1} = 1 - 0.368 = 0.632$.
    Thus η can be found from the time co-ordinate of the point where the horizontal line $F = 0.632$ intersects the straight line through the data points. From Figure 2.5,

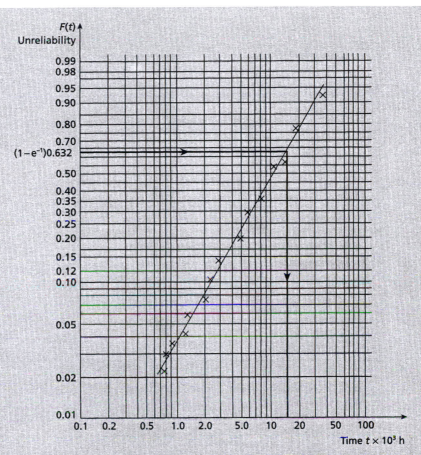

Figure 2.5 Unreliability data plotted on Weibull probability graph paper (Example 2.1)

$\eta = 15 \times 10^3$ **hours**. The straight line passes through the points (2000, 0.08) and (20 000, 0.75) and β can be estimated from the corresponding gradient, *i.e.*

$$\beta = \frac{\log_e \log_e \left(\dfrac{1}{1 - 0.75} \right) - \log_e \log_e \left(\dfrac{1}{1 - 0.08} \right)}{\log_e (20\,000) - \log_e (2000)}$$

$$= \frac{\log_e \log_e (4.00) - \log_e \log_e (1.087)}{9.90 - 7.60}$$

$$= \frac{0.327 - (-2.484)}{2.30} = \frac{2.811}{2.30} \approx \textbf{1.22}$$

(d) $R(t) = \exp\left\{-\left(\dfrac{t}{15\,000}\right)^{1.22}\right\}$

Therefore

$R(10\,000) = \exp\{-(0.667)^{1.22}\} = 0.543$ *i.e.* 54.3%

(e) $\lambda(t) = \dfrac{\beta}{\eta}\left(\dfrac{t - t_0}{\eta}\right)^{\beta-1} = \dfrac{1.22}{15 \times 10^3}\left(\dfrac{t}{15\,000}\right)^{0.22}$

$= 8.13 \times 10^{-5}\left(\dfrac{t}{15\,000}\right)^{0.22}$

i.e. $\lambda(t)$ increases gradually with time.

(f)   From Equation 2.4

$$\text{MTTF} = \int_0^\infty R(t)\,dt = \int_0^\infty \exp\left\{-\left(\dfrac{t}{15\,000}\right)^{1.22}\right\}dt$$

if we let $\tau = t/15\,000$, we have MTTF $= 15\,000 \int_0^\infty e^{-\tau^{1.22}}\,d\tau$. The integral is difficult to evaluate analytically but an approximate value $\int_0^\infty e^{-\tau^{1.22}}\,d\tau \approx 0.93$ can be found by numerical calculation of the area under $e^{-\tau^{1.22}}$ between $\tau = 0$ and $\tau = 3.5$. This gives **MTTF $\approx 15\,000 \times 0.93 \approx$ 13 950 hours.**

Summary

This chapter explained how **reliability** is defined and quantified. Concepts such as **mean time between failures, failure rate** and **availability** were introduced. The variation in reliability with time was discussed using **constant** and **Weibull** failure rate models.

Reference

1.   BS 5760 (1986) *Reliability of Constructed or Manufactured Products, Systems, Equipments and Components, Part O: Introductory Guide to Reliability*, British Standards Institution, London, p. 2.

Self-assessment questions

2.1 A reliability test was carried out on 25 non-repairable Christmas tree light bulbs. The times at which failures occurred (in units of 10^3 hours) were as follows:

0.4, 0.9, 1.3, 1.7, 1.9, 2.4, 3.0, 3.3, 3.6, 4.1, 4.5, 5.0, 5.3, 5.6, 6.1, 6.4, 6.9, 7.1, 7.5, 7.9, 8.3, 8.6, 8.9, 9.5, 9.9

(a) Use the data to estimate mean time to failure and mean failure rate.

(b) The reliability of the bulbs can be described by the equation:

$$R(t) = 1 - 10^{-4}t$$

Use the equation to estimate MTTF and $\bar{\lambda}$.

(Hint: upper limit on integral, $t = 10^4$ hours)

2.2 A batch of one hundred repairable pumps were tested over a 12-month period. Twenty failures were recorded and the corresponding down times in hours were as follows:

5, 6, 7, 8, 4, 7, 8, 10, 5, 4, 8, 5, 4, 5, 6, 5, 4, 9, 8, 6

Calculate:

(a) mean down time

(b) mean time between failures

(c) mean failure rate

(d) availability.

2.3 A gas turbine has a constant failure rate of 0.1 per year. Calculate the reliability and unreliability after:

(a) 1.0 year

(b) 2.0 years

(c) 5.0 years.

2.4 (a) Explain the general form of the variation of instantaneous failure rate with time for an engineering product.

(b) The Weibull instantaneous failure rate function is given by:

$$\lambda(t) = \frac{\beta}{\eta}\left(\frac{t - t_0}{\eta}\right)^{\beta-1}$$

(i) Explain the significance of the parameters β, η and t_0.

(ii) Explain how the above function can be used to describe the general variation in instantaneous failure rate discussed in (a).

2.5 The failure behaviour of a mechanical component can be described by a Weibull function with $t_0 = 0$, $\eta = 1.5 \times 10^4$ hrs and $\beta = 1.5$. A thousand components are placed on test at $t = 0$; calculate:

(a) the probability of survival to $t = 10^4$ hrs;
(b) the expected time at which 900 components have failed.

2.6 The reliability of spark plugs can be described by a Weibull failure rate function with $t_0 = 0$, $\eta = 4.0$ years and $\beta = 2.0$.

(a) Calculate the time at which the probability of failure is 0.25.
(b) Calculate the probability of failure after 4 years.
(c) Estimate the mean time to failure (MTTF).

$$\left(\text{Hint: } \frac{1}{\sqrt{2\pi}} \int_0^\infty \exp\left(\frac{-z^2}{2}\right) dz = 0.500 \right)$$

2.7 The reliability of 1000 components was measured over a total time interval of 10^4 hours. At the beginning of the interval all components were working to specification. Table Q2.7 shows how the cumulative number of failures n increases with time t.

(a) Find the time variation in unreliability and reliability.
(b) Estimate the values of the parameters t_0, η and β.

Table Q2.7 Cumulative failures increasing with time

Time t hours $\times 10^3$	Cumulative total no. of failures n
0	0
0.5	10
1.0	25
1.5	122
2.0	212
3.0	541
4.0	729
5.0	874
6.0	909
7.5	953
10.0	984

Reliability of systems

Introduction

In Chapter 2 we saw how reliability can be quantified in terms of failure probability, failure rate, mean time between failures, *etc*. Many branches of engineering are concerned with the development and production of **systems** which are made up of several simpler **elements** or components. The reliability of the system will depend crucially on the reliability of the elements present. This chapter shows how to calculate system reliability from individual element reliabilities and gives worked examples for a number of practical systems.

3.2 **Series systems**

Figure 3.1 shows a system of m elements in series or cascade with individual reliabilities $R_1, R_2, \ldots, R_i, \ldots, R_m$ respectively. The system will only survive if every element survives; if one element fails then the system fails. Assuming that the reliability of each element is independent of the reliability of the other elements, then the probability that the system survives is the probability that element 1 survives **and** the probability that 2 survives **and** the probability that 3 survives, *etc*. The system reliability R_{SYST} is therefore the **product** of the individual element reliabilities (Section 1.4.4), *i.e.*:

$$R_{\text{SYST}} = R_1 R_2 \ldots R_i \ldots R_m \qquad \textbf{reliability of series system} \qquad (3.1)$$

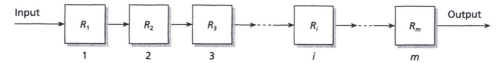

Figure 3.1   Reliability of series system

If we further assume that each of the elements can be described by a constant failure rate λ (Section 2.6.1), and if λ_i is the failure rate of the ith element, then R_i is given by the exponential relation (Equation 2.16)

$$R_i = e^{-\lambda_i t} \tag{3.2}$$

Thus

$$R_{\text{SYST}} = e^{-\lambda_1 t} e^{-\lambda_2 t} \ldots e^{-\lambda_i t} \ldots e^{-\lambda_m t} \tag{3.3}$$

so that if λ_{SYST} is the overall system failure rate:

$$R_{\text{SYST}} = e^{-\lambda_{\text{SYST}} t} = e^{-(\lambda_1 + \lambda_2 + \ldots + \lambda_i + \ldots + \lambda_m)t} \tag{3.4}$$

and

$$\boxed{\lambda_{\text{SYST}} = \lambda_1 + \lambda_2 + \ldots + \lambda_i + \ldots + \lambda_m} \quad \begin{array}{l}\textbf{failure rate of system}\\ \textbf{of } \textit{\textbf{m}} \textbf{ elements in series}\end{array} \tag{3.5}$$

This means that the overall failure rate for a series system is the **sum** of the individual element or component failure rates. Equations 3.1 and 3.5 show the importance of keeping the number of elements in a series system to a minimum; if this is done the system failure rate will be minimum and the reliability maximum.

Protective systems (Chapter 7) are characterized by having element and system **unreliabilities** F that are very small. The corresponding element and system reliabilities R are therefore very close to 1; for example, 0.9999 may be typical. In this situation, the calculation of R_{SYST} using Equation 3.1 may be arithmetically unwieldy and an alternative equation involving unreliabilities may be more useful. Since $R_{\text{SYST}} = 1 - F_{\text{SYST}}$ and $R_i = 1 - F_i$, Equation 3.1 becomes:

$$\begin{aligned} 1 - F_{\text{SYST}} &= (1 - F_1)(1 - F_2) \ldots (1 - F_i) \ldots (1 - F_m) \\ &= 1 - (F_1 + F_2 + \ldots + F_i + \ldots + F_m) \\ &\quad + \text{terms involving products of the } F\text{'s} \end{aligned} \tag{3.6}$$

If the **individual** F_i **are small, i.e.** $F_i \ll 1$, the terms involving the products of Fs can be neglected giving the **approximate** equation:

$$\boxed{F_{\text{SYST}} \approx F_1 + F_2 + \ldots + F_i + \ldots + F_m} \quad \begin{array}{l}\textbf{unreliability of series}\\ \textbf{system with small } \textit{\textbf{F}}\textbf{'s}\end{array} \tag{3.7}$$

i.e. the system unreliability is approximately the sum of the element unreliabilities.

3.3 | Parallel systems

Figure 3.2 shows an overall system consisting of n individual elements or systems in parallel with individual unreliabilities $F_1, F_2, \ldots, F_j, \ldots, F_n$ respectively. All of the elements or systems are expected to be working normally, *i.e.* to be active, and each one is capable of meeting the functional requirements placed on the overall system. However, only one element/system is necessary to meet these requirements; the remainder increase the reliability of the overall system. This is termed **active redundancy**. The overall system will only fail if every element/system fails; if one element/system survives the overall system survives. Assuming that the reliability of each element/system is independent of the reliability of the other elements, then the probability that the overall system fails is the probability that element/system 1 fails **and** the probability that 2 fails **and** the probability that 3 fails, *etc.* The overall system unreliability F_{SYST} is therefore the **product** of the individual element system unreliabilities (Section 1.4.4), *i.e.*

$$F_{\text{SYST}} = F_1 F_2 \ldots F_j \ldots F_n \qquad \begin{array}{l}\textbf{unreliability of} \\ \textbf{parallel system}\end{array} \qquad (3.8)$$

Comparing Equations 3.1 and 3.8, we see that for **series systems**, system **reliability** is the product of element **reliabilities**, whereas for **parallel** systems, system **unreliability** is the product of element **unreliabilities**. Often the individual elements/systems are identical, so that $F_1 = F_2 = \ldots = F_i = \ldots = F_n = F$; this gives:

$$F_{\text{SYST}} = F^n \qquad (3.9)$$

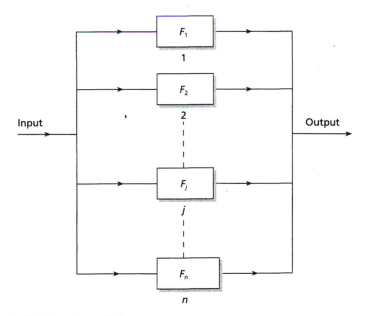

Figure 3.2   Reliability of parallel system

Thus if $F = 0.1$ and there are four channels in parallel we have $F_{SYST} = 10^{-4}$. We see, therefore, that increasing the number of individual elements/systems in a parallel system increases the overall system reliability.

3.4 General series–parallel system

Figure 3.3 shows a general series–parallel system; this consists of n identical subsystems in parallel, and each subsystem consists of m elements in series. If R_{ji} is the reliability of the ith element in the jth subsystem, then from Equation 3.1, the reliability of the jth subsystem is:

$$R_j = R_{j1}R_{j2} \dots R_{ji} \dots R_{jm} = \prod_{i=1}^{i=m} R_{ji} \tag{3.10}$$

The corresponding unreliability of the jth subsystem is:

$$F_j = 1 - \prod_{i=1}^{i=m} R_{ji} \tag{3.11}$$

From Equation 3.8 the overall system unreliability is:

$$F_{OVERALL} = \prod_{j=1}^{j=n} \left[1 - \prod_{i=1}^{i=m} R_{ji} \right] \tag{3.12}$$

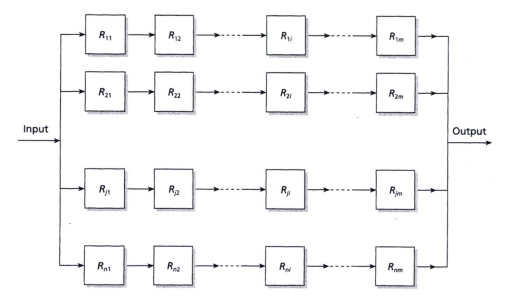

Figure 3.3 Reliability of series–parallel system

Example 3.1

A flow measurement system consists of an orifice plate ($\lambda = 0.75$), differential pressure transmitter ($\lambda = 1.0$), square root extractor ($\lambda = 0.1$) and recorder ($\lambda = 0.1$) connected in series. Calculate the probability of losing the flow measurement after 0.5 year for the following:

(a) a single flow measurement system;
(b) three identical flow measurement systems in parallel;
(c) a system with three orifice plates, three differential pressure transmitters and a single middle value selector relay ($\lambda = 0.1$). The relay selects the transmitter output signal which is neither highest nor lowest. The selected signal is then passed to a single square root extractor and recorder.
(d) What conclusions can be drawn from these calculations?

Annual failure rate data are given in brackets: assume that all systems were initially checked and found to be working correctly.

Solutions

(a) **Single system** (Figure 3.4)

System failure rate $\lambda_{\text{SYST}} = \lambda_1 + \lambda_2 + \lambda_3 + \lambda_4 = 1.95$

Unreliability $\quad F_{\text{SYST}}(t) = 1 - e^{-\lambda_{\text{SYST}}t}$
$$= 1 - e^{-1.95 \times 0.5} = 1 - e^{-0.975}$$
$$= 1 - 0.377 = 0.623$$

Probability of failure after 0.5 year = **0.623**

(b) **Three identical systems in parallel**

Overall unreliability $F_{\text{OVERALL}} = F_{\text{SYST}}^3$

Probability of failure after 0.5 year = $(0.623)^3 = $ **0.242**

(c) **System with middle value selector relay** (Figure 3.5)

Combined failure rate of O/P + D/P = 1.75

Probability of failure after 0.5 year = $1 - e^{-1.75 \times 0.5}$
$$= 1 - e^{-0.875} = 1 - 0.417$$
$$= 0.583$$

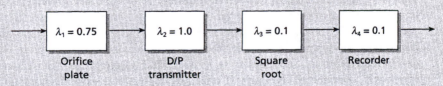

Figure 3.4 Single flow measurement system

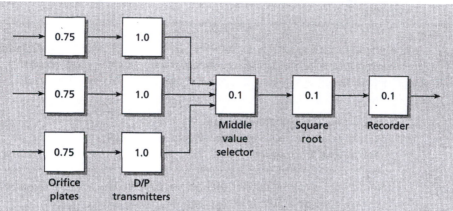

Figure 3.5   System with middle value selector relay

Selector selects one channel from three possible O/P + D/P channels. Overall unreliability of O/P + D/P channels = $(0.583)^3 = $ **0.198**

i.e.

Reliability of OP/DP channels = $1 - 0.198 = 0.802$

Combined failure rate of the remaining three elements = $0.1 + 0.1 + 0.1 = 0.3$

Reliability of MVS + square root + recorder = $e^{-0.3 \times 0.5}$

$$= 0.861$$

Overall system reliability $R_{OVERALL} = 0.802 \times 0.861$

$$= 0.690$$

System unreliability $F_{OVERALL} = $ **0.310**

(d) **Conclusions** Single system (a) has the highest unreliability; three systems in parallel (b) have the lowest unreliability but are expensive. System (c) is slightly more unreliable than (b) but is significantly cheaper.

3.5 | Majority parallel systems

This is another example of **active redundancy**. Here there are n elements or systems in parallel, all of which are expected to be working normally. The overall system will continue to function correctly if only m $(m \leqslant n)$ of the n elements/systems are working normally; the remaining $(n - m)$ elements/systems ensure extra reliability. Such a system is referred to as 'm out of n' or m oo n. One example is a four-engined aircraft which will continue to fly safely if only two engines are working normally; this is a 2 oo 4 system. An example of a 3 oo 4 system is a boiler feedwater system with four pumps normally working; each pump can supply up to one-third of the required total flow rate so that the system will work with only three pumps.

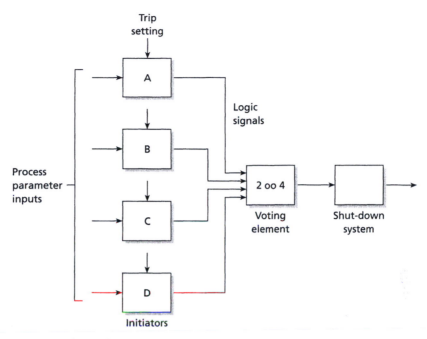

Figure 3.6   Majority voting system

Majority voting systems are used to protect hazardous plant and processes and have applications in the chemical, nuclear and aerospace industries. Figure 3.6 shows a typical system; here four identical initiators A, B, C, D give 0 or 1 logic signals depending on whether a process parameter, such as temperature or pressure, is below or above the trip setting. These logic signals are passed to a 2 out of 4 voting element; if 2 or more initiators are in a trip condition then the voting element sends a trip signal to the shut-down system. Suppose that R and F are, respectively, the reliability and unreliability of the individual initiators. The overall initiation system fails to protect the plant if **either** all 4 initiators fail **or** if any 3 initiators fail. If 2 or fewer initiators fail, there are still sufficient left to trip the plant. Since the Fs are normally small, the rare events approximation (Section 1.4.5) is valid here and the overall system unreliability is therefore the **sum** of the following probabilities:

F_{INIT} = Probability that A **and** B **and** C **and** D fail

  +

Probability that A **and** B **and** C fail

  +

Probability that B **and** C **and** D fail

  +

Probability that A **and** C **and** D fail

  +

Probability that A **and** B **and** D fail

From Section 1.4.4, each of the above terms is the **product** of individual unreliabilities and reliabilities; this gives:

$$F_{\text{INIT}} = F^4 + F^3R + F^3R + F^3R + F^3R$$
$$= F^4 + 4F^3R$$
$$= F^3(F + 4R) \tag{3.13}$$

The above result can also be obtained from the binomial expansion of $(F + R)^4$; we have:

$$(F + R)^4 = F^4 + 4F^3R + 6F^2R^2 + 4FR^3 + R^4 \tag{3.14}$$

The first term F^4 again represents the probability of all four initiators failing and the second term $4F^3R$ the total probability of three initiators failing. If R is very close to 1 and F a lot smaller than 1, Equation 3.13 reduces to:

$$F_{\text{INIT}} \approx 4F^3 \tag{3.15}$$

The unreliability of the complete protective system consisting of initiating, voting and shut-down elements in series can be calculated using either Equation 3.1 or 3.7; if the Fs are small, the approximate Equation 3.7 can be used giving:

$$F_{\text{SYST}} \approx F_{\text{INIT}} + F_{\text{VOTING}} + F_{\text{SHUT-DOWN}} \tag{3.16}$$

Further examples of majority voting systems are given in Chapter 7. An example on the use of the binomial expansion in a majority parallel system follows.

Example 3.2

A decision has to be made whether to buy two, three or four diesel generators for the electrical power system for an oil platform. Each generator will normally be working and can supply up to 50% of the total power demand. The reliability of each generator can be specified by a constant failure rate of 0.21 per year. The generators are to be simultaneously tested and proved at six-monthly intervals.

(a)   The required system reliability must be at least 0.99. Use the binomial expansion $(R + F)^n$ to decide how many generators must be bought.
(b)   Calculate the MTTF for the chosen system.
(c)   Explain how the reliability of a system with three generators is affected by having the three six-monthly tests staggered by two months.

Solutions

(a)   For a single generator:
Reliability just before test: $R = e^{-\lambda T} = e^{-0.21 \times 0.5} = 0.9$
Unreliability $F = 1 - R = 0.1$.

Since each generator can supply 50% of the power demand there must be at least two working, *i.e.* we have a 2 oo n system.

n = 2

$(R + F)^2 = R^2$ (Probability of two working = $(0.9)^2 = \mathbf{0.81}$)
$\qquad\qquad + 2RF$ (Probability of one working)
$\qquad\qquad + F^2$ (Probability of none working)

n = 3

$(R + F)^3 = R^3$ (Probability of three working = $(0.9)^3 = 0.729$)
$\qquad\qquad + 3R^2F$ (Probability of two working = $3 \times 0.9^2 \times 0.1 = 0.243$)
$\qquad\qquad + 3RF^2$ (Probability of one working)
$\qquad\qquad + F^3$ (Probability of none working)

Total probability of at least two working = **0.972**

n = 4

$(R + F)^4 = R^4$ (Probability of four working = $(0.9)^4 = 0.6561$)
$\qquad\qquad + 4R^3F$ (Probability of three working = $4 \times (0.9)^3 \times 0.1 = 0.2916$)
$\qquad\qquad + 6R^2F^2$ (Probability of two working = $6 \times (0.9)^2 \times 0.1^2 = 0.0486$)
$\qquad\qquad + 4RF^3$ (Probability of one working)
$\qquad\qquad + F^4$ (Probability of none working)

Total probability of at least two working = **0.996**

Since system reliability must be at least 0.99, the oil platform needs to install four generators.

(b) Probability that the system survives = probability that at least two generators survive. Thus for $n = 4$

$$R_{\mathrm{SYST}} = R^4 + 4R^3F + 6R^2F^2$$

Since

$$R = e^{-\lambda t} \quad \text{and} \quad F = 1 - e^{-\lambda t}$$
$$R_{\mathrm{SYST}} = e^{-4\lambda t} + 4e^{-3\lambda t}(1 - e^{-\lambda t}) + 6e^{-2\lambda t}(1 - e^{-\lambda t})^2$$
$$= 3e^{-4\lambda t} - 8e^{-3\lambda t} + 6e^{-2\lambda t}$$

$$\mathrm{MTTF} = \int_0^\infty R_{\mathrm{SYST}}\, dt$$

$$= \int_0^\infty (3e^{-4\lambda t} - 8e^{-3\lambda t} + 6e^{-2\lambda t})\, dt$$

$$= 0 - \left[-\frac{3}{4\lambda} + \frac{8}{3\lambda} - \frac{3}{\lambda} \right] = \frac{13}{12\lambda}$$

MTTF = 5.16 years

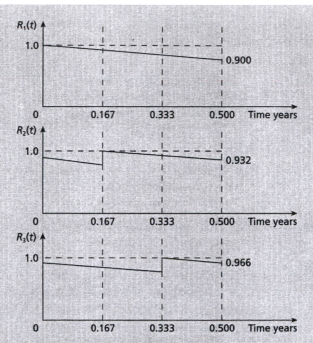

Figure 3.7 Staggered testing of three generators

(c) Staggered testing of three generators (Figure 3.7)

Generator 1 tested at $t = 0.0$

Reliability at $t = 0.5$, $R_1 = e^{-0.21 \times 0.5} = 0.900$
$$F_1 = 0.100$$

Generator 2 tested at $t = 0.167$

Reliability at $t = 0.5$, $R_2 = e^{-0.21 \times 0.333} = 0.932$
$$F_2 = 0.068$$

Generator 3 tested at $t = 0.333$

Reliability at $t = 0.5$, $R_3 = e^{-0.21 \times 0.167} = 0.966$
$$F_3 = 0.034$$

Probability of three working $= R_1 R_2 R_3 = (0.900)(0.932)(0.966)$
$$= 0.810$$

Probability of two working $= R_1 R_2 F_3 + R_1 R_3 F_2 + R_2 R_3 F_1$
$$= (0.900)(0.932)(0.034) + (0.900)(0.966)(0.068)$$
$$+ (0.932)(0.966)(0.100)$$
$$= 0.0285 + 0.0591 + 0.0900 = 0.178$$

Therefore system reliability = Probability of at least two working
= **0.988**

Thus with staggered testing, the reliability of the three-generator system is increased from 0.972 to 0.988 but will still have to install four generators to achieve better than 0.99.

3.6 Standby redundancy

We saw in Sections 3.3 and 3.5 how the use of n elements or subsystems in parallel increases the reliability of the overall system. These systems involved **active redundancy** where normally all of the elements/subsystems are in continuous operation. In some critical systems, for example boiler feedwater pumps for a chemical plant or emergency diesel-powered electrical generators for a nuclear reactor, it is essential to have more than one unit but it may be difficult or uneconomic to run all of the items continuously. In these situations **standby redundancy** is used. Here only one unit is operating at a time; the other units are shut down and are only brought into operation when the operating unit fails.

Figure 3.8 shows a general standby system consisting of n identical units, each with failure rate λ and a switching system S. Normally only unit 1 is operating and the others are shut down. If unit 1 fails then unit 2 is switched in, if unit 2 fails unit 3 is switched in, and so on until unit n is switched in.

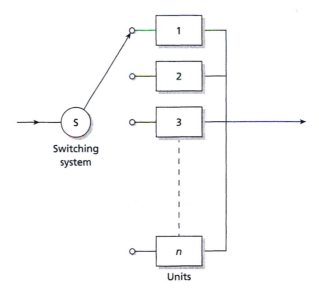

Figure 3.8 Standby system

Assuming the switching system has perfect reliability (*i.e.* $R_S = 1$), then the reliability of the standby system can be shown to be given by the **Cumulative Poisson** distribution (Appendix A):

$$R(t) = \exp(-\lambda t) \sum_{k=0}^{n-1} \frac{(\lambda t)^k}{k!}$$ **reliability of standby system**   (3.17)

Thus for $n = 1$, we have as expected:

$$R(t) = \exp(-\lambda t)$$

For $n = 2$, $R(t)$ is increased to:

$$R(t) = \exp(-\lambda t)[1 + \lambda t]$$

The term $\lambda t e^{-\lambda t}$ represents the increase in reliability due to adding one standby unit.
For $n = 3$, $R(t)$ is further increased to:

$$R(t) = \exp(-\lambda t)[1 + \lambda t + \tfrac{1}{2}(\lambda t)^2]$$

The term $\tfrac{1}{2}(\lambda t)^2 e^{-\lambda t}$ represents the further increase in reliability due adding a second standby unit.

If we take into account the imperfect reliability of the switching system then these improvements are reduced.

Example 3.3

A standby boiler feedwater system consists of n identical pumps and a flow diverter. The diverter enables the discharge from any pump to be connected to the output pipe so that the system is run with one pump operational and the remainder on standby. The overall system is to have a reliability better than 0.97 over a period of six months. Use the data given below to find the value of n.

Constant failure rate for a single pump = 0.45 per year
Reliability of a single diverter operation = 0.900

Solutions

(a) **$n = 1$ – one operational pump**

$R(t) = e^{-\lambda t} = e^{-0.45 \cdot 0.5} = 0.7985$ *i.e.* reliability is too low

(b) **$n = 2$ – pump 1 operational, pump 2 standby**

Overall reliability = Probability that operational pump 1 is working
+
Probability that standby pump 2 is working AND probability of one successful diversion operation

$$R(t) = e^{-\lambda t} + \lambda t\, e^{-\lambda t} R_D$$
$$= 0.7985 + (0.45)(0.5)(0.7985)(0.900)$$
$$= \mathbf{0.9602}\ i.e.\ \text{reliability is too low}$$

(c) **$n = 3$ – pump 1 operational, pumps 2 & 3 standby**

Overall reliability = Probability that operational pump 1 is working

+

Probability that standby pump 2 is working AND
probability of one successful diversion operation

+

Probability that standby pump 3 is working AND
probability of two successful diversion operations

$$R(t) = e^{-\lambda t} + \lambda t e^{-\lambda t} R_D + 1/2\ (\lambda t)^2 e^{-\lambda t} R_D{}^2$$
$$= 0.9602 + 0.5(0.45 \times 0.5)^2 \cdot 0.7985 \cdot (0.900)^2$$
$$= 0.9602 + 0.0164 = \mathbf{0.9766}\ i.e.\ \text{reliability meets specification}$$

3.7 | Common mode failures

So far in this chapter we have assumed that the reliability of a given element in a system is independent of the reliability of the other elements in the system. In practice, however, there may be a single fault which causes either:

(i)   elements of a given type in each of several systems to fail, or
(ii) several elements in a single system to fail simultaneously.

This type of failure is called **common mode failure**. The reliability of the element is made up of two components, an **independent reliability** R_I which is independent of the other elements and a **common mode reliability** R_{CMS}. The element survives if both of these components survive; the overall element reliability is therefore the product of these components, *i.e.*:

$$R = R_I \cdot R_{CMS} \tag{3.18}$$

Common mode failures are due to a variety of causes.

(a) Incorrect design of a given element. An example is an incorrectly designed sealing system which causes all pumps of a given type to fail prematurely.
(b) Defective or inappropriate materials/components used in element manufacture. For example, iron-constantan thermocouples used to measure the temperature of sulphuric acid will corrode and fail prematurely.
(c) Faults in the manufacturing process. If a boring machine is wrongly set up then engine cylinders of a given type may have too small a diameter. This will mean insufficient piston-cylinder clearance and excessive wear.

(d)   Incorrect installation, operation and maintenance procedures. For example, if the wrong lubricant is used with a given type of engine then excessive wear will result.

(e)   Several elements in a system are dependent for their operation on a single common element; if this element fails then all of the elements fail.

One frequent example of (e) is an electronic system where several of the constituent circuits share a common electrical power supply; failure of the power supply causes all of these circuits to fail. This is shown in Figure 3.9(a). The overall system can be regarded as consisting of the **functional system** with reliability R_{FS} and the **common mode system** with reliability R_{CMS}. The probability that the overall system survives is the probability that the functional system survives **and** the probability that the common mode system survives. The overall system reliability R_S is therefore the product of R_{FS} and R_{CMS}, *i.e.*:

$$R_S = R_{FS}R_{CMS} \tag{3.19}$$

There are several possible ways of increasing the common mode reliability R_{CMS}. R_{CMS} can be increased by having a **redundant** common mode system. Figure 3.9(b) shows an electronic functional system with a standby common mode system consisting of two mains-driven DC power supplies and a switch. However, the reliability of this system is limited by the reliability of the 240 V mains supply which affects both power supplies; a better system would have one mains power supply and one battery-operated power supply.

A better general approach is to use **equipment diversity**: here we have more than one functional system, where each system carries out exactly the same function, but is made up of different elements with different operating principles. Figure 3.9(c) shows an overall system with the same function being performed by two diverse systems 1 and 2, with corresponding functional reliabilities R_{FS1}, R_{FS2} respectively and corresponding common mode reliabilities R_{CMS1}, R_{CMS2} respectively. The overall system therefore can be reduced to a system with unreliability $(1 - R_{FS1}R_{CMS1})$ in parallel with a system with unreliability $(1 - R_{FS2}R_{CMS2})$. Using Equation 3.8 the overall system reliability R_S is given by:

$$(1 - R_S) = (1 - R_{FS1}R_{CMS1})(1 - R_{FS2}R_{CMS2})$$

i.e.

$$R_S = R_{FS1}R_{CMS1} + R_{FS2}R_{CMS2} - R_{FS1}R_{FS2}R_{CMS1}R_{CMS2} \tag{3.20}$$

We see, therefore, that if there is a common mode failure in system 2, *i.e.* R_{CMS2} falls to zero, then system 1 continues to operate and the overall system reliability is now equal to $R_{FS1}R_{CMS1}$, *i.e.* that of system 1. An example of equipment diversity is an overall temperature measurement system consisting of two systems; one electronic with an electrical power supply, the other pneumatic with a compressed air supply. Here, if the electrical mains supply fails then the pneumatic system survives; if the compressed air supply fails then the electronic system survives.

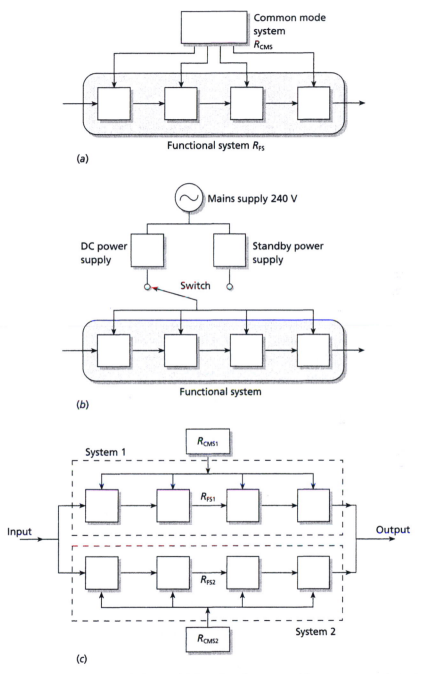

Figure 3.9 Common mode reliability: (a) functional system with common mode system; (b) functional system with standby common mode system; (c) systems with equipment diversity

The reliability of this **diverse** system should be contrasted with that of a **redundant** system consisting of two identical systems in parallel each with functional reliability R_{FS} and common mode reliability R_{CMS}. Using Equation 3.20 the overall system reliability is given by:

$$R_S = 2R_{FS}R_{CMS} - R_{FS}^2 R_{CMS}^2 \qquad (3.21)$$

If $R_{CMS} = 1$, $R_S = R_{FS}(2 - R_{FS})$ and the overall system is less susceptible to failure of a functional system than a single functional system. However, the overall system is susceptible to a fault which causes both of the identical common mode systems to fail; if $R_{CMS} = 0$, $R_S = 0$. This corresponds to the situation where both temperature measurement systems are electronic with mains-driven DC power supplies and the overall system is again vulnerable to the loss of mains supply.

To have complete **equipment diversity**, the functional systems should not only consist of different equipment with different operating principles but should also, where appropriate:

(a)   have separate locations
(b)   have separate signal routes
(c)   be made from different materials
(d)   be tested/repaired at different times
(e)   be tested/repaired by different technicians.

In Section 3.3 we saw that if n identical functional elements or systems, each with unreliability F are connected in **parallel**, the overall system unreliability F_{SYST} is reduced to $F_{SYST} = F^n$. However, this reliability improvement will not occur if all of the functional systems are reliant on a single common mode system. Figure 3.10(a) shows an overall system of three functional systems each with reliability R_{FS} and a single common mode system with reliability R_{CMS}. The overall functional system has an unreliability of $(1 - R_{FS})^3$ and a reliability of $1 - (1 - R_{FS})^3$. The overall system will only survive if the functional system **and** the common mode system survive, so that the overall system reliability is given by:

$$R_S = R_{CMS}[1 - (1 - R_{FS})^3] \qquad (3.22)$$

R_S thus depends on the reliability of the single common mode system and the reliability improvement of using a parallel system is lost. The reliability is improved if each functional system has its own associated common mode system (*e.g.* power supply) as shown in Figure 3.10(b). The overall system reliability is now increased to:

$$R_S = [1 - (1 - R_{FS}R_{CMS})^3] \qquad (3.23)$$

A further reliability improvement can be made by using three diverse systems as explained earlier.

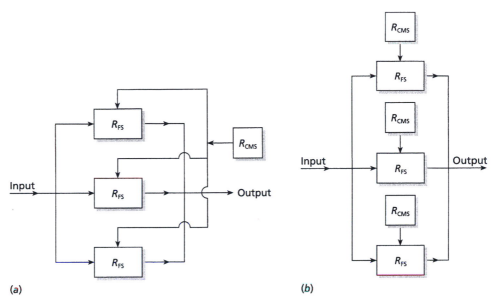

Figure 3.10 Common mode reliability in redundant systems: (a) three functional systems with a single common mode system; (b) three functional systems each with an associated common mode system

3.8 | Availability of systems with repair

Many engineering systems contain elements or subsystems that can be repaired. In the systems discussed so far in this chapter, the assumption has been made that repair activity takes place when all elements or subsystems have failed. However, repair of failed elements that are **down** can take place simultaneously with the normal operation of working elements that are **up**. This section examines the **availability** of systems with simultaneous repair.

In Section 2.3.2 we examined the failure pattern for N items of a single repairable element over a test interval T and defined **mean time between failures MTBF** and **mean down time MDT** (Equations 2.6 and 2.5). We then defined availability A by (Section 2.4):

$$\text{availability} = \frac{\text{Total up time}}{\text{Total up time} + \text{Total down time}}$$

giving:

$$A = \frac{\text{MTBF}}{\text{MTBF} + \text{MDT}}$$

(3.24)

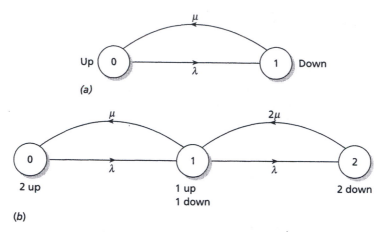

Figure 3.11   Markov state diagrams: (a) two-state diagram for single element; (b) three-state diagram for standby system

Using similar arguments, unavailability U is given by:

$$U = \frac{\text{MDT}}{\text{MTBF} + \text{MDT}}$$

(3.25)

so that we have $A + U = 1$. We see from Equations 3.24 and 3.25 that A and U depend on both reliability (MTBF) and maintainability (MDT).

In order to be able to calculate the availability of systems of repairable elements the techniques of **Markov analysis** can be used.[1] We first apply this analysis to a single repairable element; at any time t the element can only be in one of two possible states: state 0 the **up** state, or state 1 the **down** state (Figure 3.11(a)). The element moves from state 0 to state 1 as a result of failure; a constant failure rate λ is assumed, where $\lambda = 1/\text{MTBF}$. Similarly the element moves from state 1 to state 0 as a result of repair; a constant repair rate μ is assumed, where $\mu = 1/\text{MDT}$. If $P_0(t)$, $P_1(t)$ are the probabilities of the element being in states 0 and 1 respectively at time t, then the probability that the element is in state 0 at time $t + \Delta t$ is given by:

$P_0(t + \Delta t) =$ Probability of being in 0 at time t **and**
  probability of not failing between t and $t + \Delta t$
  or Probability of being in 1 at time t **and**
  probability of being repaired between t and $t + \Delta t$.

From Equation 2.11 the probability of failure between t and $t + \Delta t$ is $\lambda\Delta t$, and the probability of not failing is $(1 - \lambda\Delta t)$; similarly the probability of repair is $\mu\Delta t$. Using the addition and multiplication rules for probabilities (Equations 1.26 and 1.29) we have:

$P_0(t + \Delta t) = P_0(t)(1 - \lambda\Delta t) + P_1(t)\mu\Delta t$

(3.26)

Using similar arguments for state 1 we have:

$$P_1(t + \Delta t) = P_0(t)\lambda\Delta t + P_1(t)(1 - \mu\Delta t) \tag{3.27}$$

Rearranging these equations gives:

$$\frac{P_0(t + \Delta t) - P_0(t)}{\Delta t} = -\lambda P_0(t) + \mu P_1(t) \tag{3.28}$$

$$\frac{P_1(t + \Delta t) - P_1(t)}{\Delta t} = \lambda P_0(t) - \mu P_1(t) \tag{3.29}$$

In the limit that $\Delta t \rightarrow 0$, the equations become:

$$\frac{dP_0}{dt} = -\lambda P_0(t) + \mu P_1(t) \tag{3.30}$$

$$\frac{dP_1}{dt} = \lambda P_0(t) - \mu P_1(t) \tag{3.31}$$

Since $P_0(t) + P_1(t) = 1$, Equation 3.30 simplifies to:

$$\frac{dP_0}{dt} = -(\mu + \lambda)P_0(t) + \mu \tag{3.32}$$

Integrating Equation 3.32 and using the initial condition $P_0(0) = 1$ gives the following solution:

$$P_0(t) = \frac{\mu}{\mu + \lambda} + \frac{\lambda}{\mu + \lambda}\exp[-(\mu + \lambda)t] \tag{3.33}$$

Similarly integrating Equation 3.31 gives

$$P_1(t) = \frac{\lambda}{\mu + \lambda}\{1 - \exp[-(\mu + \lambda)t]\} \tag{3.34}$$

The **steady state availability** A_{SS} is the long-term probability that the element is in the **up** state, *i.e.*:

$$A_{SS} = \lim_{t\to\infty} P_0(t) = P_0(\infty) = \frac{\mu}{\mu + \lambda} \tag{3.35}$$

Similarly the **steady state unavailability** U_{SS} is the long-term probability that the element is in the **down** state, *i.e.*:

$$U_{SS} = \lim_{t\to\infty} P_1(t) = P_1(\infty) = \frac{\lambda}{\mu + \lambda} \tag{3.36}$$

By substituting $\lambda = 1/\text{MTBF}$ and $\mu = 1/\text{MDT}$, we can show that Equations 3.35 and 3.36 are identical with Equations 3.24 and 3.25. Thus if we consider a single element with $\lambda = 1.0$ year^{-1} and MDT = 1 week = 1/52 year, *i.e.* $\mu = 52$ year^{-1}, the corresponding availability $A = 52/53 = 0.981$.

Since availability is simply the long-term probability that a repairable system is in the **up** state, the calculation of availability of series and parallel systems is similar to that of reliability. Thus for a **series system** consisting of m elements (Figure 3.1) with individual availabilities $A_1, A_2, \ldots, A_i, \ldots, A_m$ respectively, the system availability A_{SYST} is the **product** of the individual element availabilities, *i.e.*:

$$A_{\text{SYST}} = A_1 A_2 A_3 \ldots A_i \ldots A_m \quad \textbf{availability of series system} \quad (3.37)$$

Thus the availability of a system of two of the above elements in series is reduced to $A_{\text{SYST}} = (0.981)^2 = 0.963$.

Many elements are characterized by values of A very close to 1, *e.g.* 0.9999, and correspondingly small values of U, *e.g.* 0.0001. If in a series system, all the individual U_i are small, *i.e.* $U_i \ll 1$, then the system unavailability is approximately the sum of the element unavailabilities (see Equation 3.7), *i.e.*:

$$U_{\text{SYST}} \approx U_1 + U_2 + \ldots + U_i \ldots + U_m \quad \begin{array}{l}\textbf{unavailability of series} \\ \textbf{system with small } U\text{'s}\end{array} \quad (3.38)$$

For a **parallel system** of n elements (Figure 3.2) with individual unavailabilities $U_1, U_2, \ldots, U_j, \ldots, U_n$ respectively, the system unavailability is given by the **product** of the individual element unavailabilities, *i.e.*:

$$U_{\text{SYST}} = U_1 U_2 U_3 \ldots U_j \ldots U_n \quad \textbf{unavailability of parallel system} \quad (3.39)$$

The unavailability of the above element is $U = 1/53 = 0.019$; the unavailability of two elements in parallel is $U_{\text{SYST}} = (0.019)^2 = 0.000\,36$. The corresponding availability is $A_{\text{SYST}} = 0.999\,64$, *i.e.* the availability of the parallel system is therefore greater than that of the single element.

For a majority parallel or 'm out of n' system (Section 3.5), the availability can be calculated using the binomial expansion of $(A + U)^n$. Thus in the case of '2 out of 4' we have:

$$(A + U)^4 = A^4 + 4A^3U + 6A^2U^2 + 4AU^3 + U^4 \quad (3.40)$$

The A^4 term represents the availability of four units, the $4A^3U$ term the availability of three units, the $6A^2U^2$ term the availability of two units and so on. Thus the availability of two or more units is:

$$A_{\text{SYST}} = A^4 + 4A^3U + 6A^2U^2 \quad (3.41)$$

We now consider the more complex example of a **standby** system of two pumps, one operational and one on standby; if either pump fails it is repaired, if both pumps fail they are repaired simultaneously. The system therefore has three states (Figure 3.11(b)):

0 – both pumps **up** – probability $P_0(t)$
1 – one pump **up**, one pump **down** – probability $P_1(t)$
2 – both pumps **down** – probability $P_2(t)$

Using similar arguments to the two-state system discussed above, the corresponding probabilities that the system is in states 0, 1 and 2 at time $t + \Delta t$ are:

$$P_0(t + \Delta t) = P_0(t)(1 - \lambda\Delta t) + P_1(t)\mu\Delta t \tag{3.42}$$

$$P_1(t + \Delta t) = P_0(t)\lambda\Delta t + P_1(t)(1 - \lambda\Delta t)(1 - \mu\Delta t) + P_2(t)2\mu\Delta t \tag{3.43}$$

$$P_2(t + \Delta t) = P_1(t)\lambda\Delta t + P_2(t)(1 - 2\mu\Delta t) \tag{3.44}$$

For each equation $[P(t + \Delta t) - P(t)]/\Delta t$ is equal to the derivative $\mathrm{d}P/\mathrm{d}t$ in the limit $\Delta t \to 0$; this gives:

$$\frac{\mathrm{d}P_0}{\mathrm{d}t} = -\lambda P_0(t) + \mu P_1(t) \tag{3.45}$$

$$\frac{\mathrm{d}P_1}{\mathrm{d}t} = \lambda P_0(t) - (\lambda + \mu)P_1(t) + 2\mu P_2(t) \tag{3.46}$$

$$\frac{\mathrm{d}P_2}{\mathrm{d}t} = \lambda P_1(t) - 2\mu P_2(t) \tag{3.47}$$

In order to calculate the steady-state availability, we require the probabilities $P_0(\infty)$, $P_1(\infty)$, $P_2(\infty)$ corresponding to the limit that $t \to \infty$. In this limit these probabilities are constant with time, so that the corresponding derivatives are zero and the above equations simplify to:

$$0 = -\lambda P_0(\infty) + \mu P_1(\infty) \tag{3.48}$$

$$0 = \lambda P_0(\infty) - (\mu + \lambda)P_1(\infty) + 2\mu P_2(\infty) \tag{3.49}$$

$$0 = \lambda P_1(\infty) - 2\mu P_2(\infty) \tag{3.50}$$

From Equation 3.48, $P_1(\infty) = \lambda/\mu P_0(\infty)$ and from Equation 3.50,

$$P_2(\infty) = \frac{\lambda}{2\mu} P_1(\infty) = \frac{\lambda^2}{2\mu^2} P_0(\infty)$$

Since

$$P_0(\infty) + P_1(\infty) + P_2(\infty) = 1$$

then

$$\left[1 + \frac{\lambda}{\mu} + \frac{\lambda^2}{2\mu^2}\right]P_0(\infty) = 1$$

i.e.

$$P_0(\infty) = \frac{2\mu^2}{2\mu^2 + 2\mu\lambda + \lambda^2} \tag{3.51}$$

and

$$P_1(\infty) = \frac{2\mu\lambda}{2\mu^2 + 2\mu\lambda + \lambda^2} \tag{3.52}$$

The steady-state availability is then the long-term probability that the system is in either state 0 or 1, *i.e.*:

$$A_{SS} = P_0(\infty) + P_1(\infty) = \frac{2\mu(\mu + \lambda)}{2\mu^2 + 2\mu\lambda + \lambda^2} \tag{3.53}$$

Using the above figures of $\lambda = 1.0$ year^{-1}, $\mu = 52$ year^{-1} gives $A_{SS} = 5512/5513 = 0.999\,82$, thus the availability of this standby system is obviously greater than that of a single element but also greater than that of two elements in parallel.

3.9 Fault tolerant systems

A fault tolerant system continues to operate within specification (or to a lesser defined specification) when one or several faults are present. Figure 3.12 shows the general form of a fault tolerant system; it consists of a basic system, in which faults may occur, with the following features added:

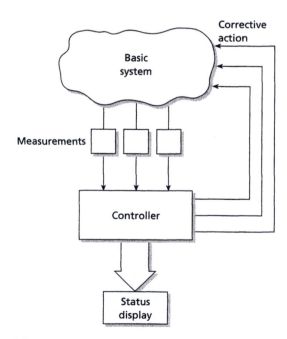

Figure 3.12   General fault tolerant system

- **Measurement** Key variables in the basic system are measured; examples are the supply voltage in an electrical circuit, the temperature of an oven in a baking process, the delivery pressure of a gas compressor. Details of sensing and signal conditioning elements used in measurement systems are given in Section 6.5.2.
- **Control** The controller is normally a computer using appropriate software. Each of the input measured values of the variables is regularly checked against the corresponding expected range. If the measured value falls outside this range then a **high** or **low** is registered as appropriate; this is done for all variables so that a pattern of highs and lows results. This pattern is then compared with a 'look up' table which gives the pattern corresponding to each possible fault condition (Section 6.3.2) so that the fault can be diagnosed. The controller then initiates corrective action appropriate to that fault.
- **Corrective action** The basic system must incorporate some form of redundancy, normally **standby redundancy** (Section 3.6). This involves redundant elements or systems which the controller can switch in. Examples are a backup battery power supply, a standby pump or microprocessor.
- **System status display** This will be an alarm message to state that a given fault has occurred and that change-over to a given standby element has taken place.

Summary

This chapter has shown how to calculate the reliability and availability of a range of practical systems given the reliability of the constituent elements and components. The following chapter gives failure rate data and models for a range of typical devices.

Reference

1. BS 5760 (1981) *Reliability of Constructed or Manufactured Products, Systems, Equipment and Components Part 2: Principles of Reliability*. British Standards Institution, London, 23–4.

Self-assessment questions

3.1 A temperature measurement system consists of a thermocouple ($\lambda = 1.1$), transmitter ($\lambda = 0.1$) and recorder ($\lambda = 0.1$) connected in series. Calculate the probability of losing the temperature measurement after six months for the following:

(a) a single temperature measurement system;
(b) three identical temperature measurement systems in parallel;
(c) a system with three thermocouples and a single middle value selector relay ($\lambda = 0.1$).

The relay selects the thermocouple output which is neither highest nor lowest. The selected signal is passed to a single transmitter and recorder. Annual failure rates are given in brackets.

3.2   An overall system consists of five subsystems in parallel. Each subsystem consists of four elements in series with annual failure rates of 0.1, 0.07, 0.05, 0.03. The overall system was initially checked and proved to be working correctly; calculate the probability of failure one year later.

3.3   A four-engined aircraft can fly safely with two or more engines working normally. Each individual engine has a constant failure rate of 10^{-4} per hour. All four engines are simultaneously tested and proved to be working correctly at intervals of 1000 hours. Find the maximum probability of overall system failure.

3.4   An electrical power system consists of identical diesel generators; each generator has a constant failure rate of 0.5 per year. Calculate the reliability of the following arrangements, three months after initial commissioning:

(a)   one operating generator;
(b)   one operating generator, one standby generator;
(c)   one operating generator, two standby generators;
(d)   one operating generator, three standby generators.

Assume perfect switching to the standby generators.

3.5   A single system can be regarded as consisting of a functional system of reliability 0.8 and a common mode system with a reliability of 0.9. Calculate the reliability of the following:

(a)   the single system;
(b)   a redundant system consisting of three identical functional systems in parallel (each with $R = 0.8$) together with a single common mode system ($R = 0.9$);
(c)   a redundant system consisting of three of the above single systems in parallel;
(d)   a diverse system consisting of three systems in parallel specified as follows:
     (i)   functional reliability = 0.8, common mode reliability = 0.9
     (ii)   functional reliability = 0.7, common mode reliability = 0.85
     (iii)   functional reliability = 0.9, common mode reliability = 0.95.

3.6   A microprocessor-based controller consists of nine different types of elements. Figure Q3.6 is a reliability block diagram for the system. The failure rate of each type of element per 10^6 hours is shown in the diagram. The mean down time for the repair of each element is eight hours.

(a)   Calculate the unavailability of each element.
(b)   Calculate the unavailability and availability for the overall system.
(c)   Estimate the effect on system availability of reducing repair mean down time to four hours.

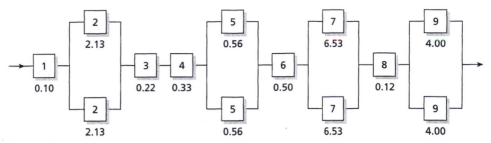

Figure Q3.6

3.7 Boiler feed water is supplied by a 3 out of 4 majority parallel active system of pumps; the availability of each pump is 0.94.

It is proposed to replace this active system by a standby system of two pumps, one operational and one on standby. If either pump fails it is repaired; if both pumps fail they are repaired simultaneously. The availability of each pump in the standby system is 0.98.

(a) Calculate the availability of the active 3 out of 4 system.
(b) Calculate the availability of the standby system.
(c) If the system down time costs are £1500 per hour, calculate down time costs for each system over a one-year period.
(d) The capital investment required to make the replacement is £500 000; this should be recovered over a three-year period by savings in down time costs. Should the proposal be accepted?

Failure rate data and models

4.1 Introduction

In the previous chapter, we saw how the reliability of a range of systems can be calculated given the reliability of the components or elements which make up the system. It is then necessary to have failure rate data for these individual components and elements. This chapter begins by presenting typical failure rate data, goes on to explain the various factors which influence failure rate and then discusses mathematical models for failure rate. The reliability of software is then discussed and the chapter concludes by examining human reliability.

4.2 Failure rate data

Failure rate was defined in Section 2.3.1, and in Section 2.5 we saw that there are several possible ways in which failure rate can vary with time. In this section failure rate data for a variety of typical engineering components and system elements are presented. A component is defined as a 'non-repairable' device, *i.e.* when it fails it is removed and thrown away; examples are a resistor, an integrated circuit, a car headlight bulb and a gasket. An element, however, is a repairable part of a system; examples are a printed circuit board, a control valve and a pump. The data are obtained by direct measurements of the frequency of failures of components/elements of a given type. This procedure is based on the assumption of constant failure rate with time. The **observed failure rate** $\bar{\lambda}$ is then given by (Section 2.3.1)

$$\overline{\lambda} = \frac{\text{Total number of failures during observation time}}{\text{Total exposure time (Total \textbf{up} time)}}$$

For non-repairable components, we can find $\overline{\lambda}$ using Equation 2.3

$$\overline{\lambda} = \frac{N_F}{\sum_{i=1}^{i=N_F} T_i} = \frac{1}{\text{MTTF}} \tag{4.1}$$

where N_F = total number of failures

$\sum_{i=1}^{i=N_F} T_i$ = total **up** time

MTTF = mean time to fail

For repairable elements, we can calculate $\overline{\lambda}$ using Equation 2.7 *i.e.*:

$$\overline{\lambda} = \frac{N_F}{NT - \sum_{j=1}^{j=N_F} T_{Dj}} = \frac{N_F}{NT - N_F \times \text{MDT}} = \frac{1}{\text{MTBF}} \tag{4.2}$$

where N_F = total number of failures

T = total observation time

N = total number of elements

$\sum_{j=1}^{j=N_F} T_{Dj}$ = total **down** time

MDT = mean **down** time

MTBF = mean time between failures

Table 4.1 gives the observed average failure rate data for typical mechanical, hydraulic and pneumatic components. The values are expressed in terms of percentage of components failing per 1000 hours, and are average values applicable to large samples of components, working at nominal stress values and operating under normal environmental conditions (see Section 4.3). These values should be divided by 100 to give failure rates in terms of the fractional number of failures per 1000 hours.

Table 4.2 gives the observed average failure rates for some electronic components expressed in terms of failure rates per 10^6 hours.[1] The data assumes low electrical stress and average environmental conditions (Section 4.3). The numerical values are for components which have been in use for over 10 000 hours, so that early failure effects are likely to be eliminated.

Table 4.3 gives the observed average failure rates for typical instruments in a variety of operating environments. These data have been taken from the UK data bank operated by the Systems Reliability Service (SRS).[1]

Table 4.4 gives the observed failure rates for larger, more complex mechanical, electrical and electronic items of equipment.[2] The failure rates are expressed in failures per 10^6 hours. In cases where there is good agreement between different sources a single figure is quoted. In most cases two or three figures are quoted to indicate a range of values.

Table 4.1 Observed average failure rate data for some mechanical, hydraulic and pneumatic components

Component	Main type of fault	Fault rate (%/kh)
Bellows (metal)	Rupture	0.5
Bourdon tubes	Leakage	0.005
	Creep	0.02
Diaphragms (metal)	Rupture	0.5
(rubber compound)	Rupture	0.8
Filters (sintered ceramic)	Blockage	0.1
	Leakage	0.1
Fulcrums	Wear of knife edge	1.0
Gaskets	Leakage	0.05
Grub screws (or clamp screws)	Loose	0.05
Guides (valve)	Jambs	0.05
Hair springs	Breakage or twisted	0.1
Hoses (plastic)	Breakage	4.0
Joints (pipe)	Breakage or leakage	0.05
Joints (mechanical)	Breakage	0.02
Joints (ball)	Breakage	0.10
Nuts, bolts, rods, shafts, etc.	Breakage or loose	0.002
O-ring seals	Leakage	0.02
Orifices	Blockage	0.05
Orifices: variable	Calibration or blockage	1.0
Pivots	Breakage or wear	0.1
Pipes (metal)	Fatigue or leakage	0.02
Pressure vessels	Rupture or leakage	0.3
Rack and pinion assemblies	Wear or jambing	0.2
Restriction	Blockage	0.5
Seals	Leakage	0.05
Springs (return force)	Breakage	0.01
Springs (calibration, helical)	Breakage	0.02
	Creep	0.2
Valves	Leakage	0.2
	Blockage	0.05

4.3 | Factors influencing failure rate

In general the failure rate of a component or element depends on four main factors:

(a) Quality
(b) Temperature
(c) Environment
(d) Stress

This section discusses the role of each of these four factors.

Table 4.2 Observed average failure rate data for some electronic components (after Wright[1])

Component	Type	Failure rate (failures/10^6 h)
Resistor	Metal oxide	0.001
	Wirewound	0.002
Capacitor	Plastic film	0.0001
	Ceramic	0.002
	Aluminium, electrolytic	0.1
	Solid tantalum, electrolytic	0.0001
Transistor	npn, small signal	0.005
FET	Small signal	0.04
Diode	Rectifier	0.004
	Logic switching	0.003
	Zener	0.02
Two-input gate	CMOS	0.02
Flip-flop	CMOS	0.1
Operational amplifier	Silicon monolithic	0.15
Voltage regulator	Silicon monolithic	0.04
Switch contact		0.1
Transformer	Audio, small signal	0.02
Fuse		0.1
Lamp	Filament (under-run)	0.05
Meter		10.0
Connector	Coaxial	0.63
Trimpot	Cermet	2.6

Table 4.3 Observed average failure rates for instruments (after Wright[1])

Instrument	Environment	Experience (item-years)	No. of failures	Failure rate (failures/y)
Chemical analyser, Oxygen	Poor, chemical/ship	4.34	30	6.92
pH meter	Poor, chemical/ship	28.08	302	10.75
Conductivity indicator	Average, industrial	7.53	18	2.39
Fire detector head	Average, industrial	1 470	128	0.09
Flow transmitter, pneumatic	Average, industrial	125	126	1.00
Level indicator, pneumatic	Average, industrial	898	201	0.22
Pressure controller	Average, industrial	40	63	1.58
Pressure indicator, dial, mechanical	Average, industrial	575	178	0.31
Pressure sensor, differential, electronic	Poor, chemical/ship	225	419	1.86
Pressure transmitter	Average, industrial	85 045	806	0.01
Recorder, pen	Average, industrial	26.02	7	0.27
Temperature indicator and alarm	Fair, laboratory	47.2	101	2.14
Temperature indicator, resistance thermomenter	Fair, laboratory	212.3	68	0.32
Temperature indicator, bimetal	Average, industrial	165	215	1.30
Temperature trip unit	Average, industrial	120	70	0.58
Thermocouple	Poor, chemical/ship	317	127	0.40
Valve, gate	Poor, chemical/ship	11 564	841	0.07
non-return	Poor, chemical/ship	1 530	101	0.07
solenoid	Poor, chemical/ship	1 804	66	0.04

Table 4.4 Observed failure rates for larger mechanical, electrical and electronic items (after Smith[2])

Item	Failure rate (in failures per million hours)		
Compressor			
Centrifugal, turbine driven	150		
Reciprocating, turbine driven	500		
Electric motor driven	100	300	
Computer			
Mainframe	4000	8000	
Mini	100	500	
Micro (CPU)	30	100	
PLC	20	50	
Generator			
AC	3	30	
DC	1	10	
Turbine set	10	200	800
Motor set	30	70	
Diesel set	125	4000	
Power supply			
DC/DC converter	2	20	
AC/DC stabilized	5	20	
Printed circuit boards			
Single-sided	0.02		
Double (plated through)	0.01	0.3	
Multilayer	0.07	0.1	
Printer (line)	300	1000	
Pumps			
Centrifugal	10	50	100
Boiler	100	700	
Fire water diesel	200	3000	
electric	200	500	
Turbine, steam	30	80	

4.3.1 Quality

The quality of a component depends on the materials from which it is made; the quality of an element again depends on the quality of the materials and components from which it is made. In general, the higher the quality of a component or element the lower the failure rate but the higher the cost. This can be illustrated by the problem of measuring the pressure at the base of a vessel containing a solution of 10% sulphuric acid in water. If the diaphragm of the pressure transducer is made from the common alloy 316 stainless steel, the acid has a moderately corrosive effect on the diaphragm so that eventually the diaphragm will fail. If, however, the diaphragm is made from the special alloy Hastelloy C, the sulphuric acid has no corrosive effect on the diaphragm. Thus a pressure transducer with a Hastelloy diaphragm has a much lower failure rate, on the sulphuric acid duty, than one with a 316 stainless steel diaphragm. The cost of the higher quality Hastelloy diaphragm is, however, much greater. Similarly an inductive displacement transducer has a higher quality, higher cost and lower failure rate than a resistive potentiometer type. The latter involves a contact sliding over a wire wound track; the contact will eventually become worn.

The quality of electronic components is specified by a quality level. One commercial specification for silicon integrated circuits defines three quality levels 1, 2 and 3 (3 being the highest, 1 the lowest). A military specification for microelectronic devices defines seven quality levels S, S-1, B, B-1, B-2, D and D-1 (S being the highest and D-1 the lowest). In all cases the higher the quality level, the lower the failure rate. The failure rate is multiplied by a **quality factor** π_Q (see Section 4.4) which decreases as the quality level increases. Thus for commercial silicon integrated circuits π_Q is 0.5, 1.0 and 2.0 for quality levels of 3, 2 and 1 respectively; for the military microelectronic devices π_Q ranges from 0.25 for S up to 20.0 for D-1. Again the higher the quality level the higher the cost; for example, a specially produced one-off electronic component for underwater applications might cost approximately £50, a mass-produced component for general application only a few pence (when purchased in bulk).

4.3.2 Temperature

The failure rate of electronic components is highly dependent on the temperature of the component. The relationship is of the form:

$$\lambda = K \exp(-E/kT) \tag{4.3}$$

where λ = component failure rate (assumed constant with time)
$\quad\quad T$ = component temperature in K
$\quad\quad E$ = activation energy of failure process in electron-volts (eV)
$\quad\quad k$ = Boltzmann's constant = 8.63×10^{-5} eV/K
$\quad\quad K$ = constant

Equation 4.3 is especially useful for calculating the failure rate λ_2 at a second temperature T_2K, given the failure rate λ_1 at a first temperature T_1K:

$$\frac{\lambda_2}{\lambda_1} = \exp\left[\frac{E}{k}\left(\frac{1}{T_1} - \frac{1}{T_2}\right)\right] \tag{4.4}$$

Figure 4.1 shows how the failure rate of a silicon NPN transistor varies with both temperature and voltage stress ratio (see Section 4.3.4). The graphs indicate how sharply

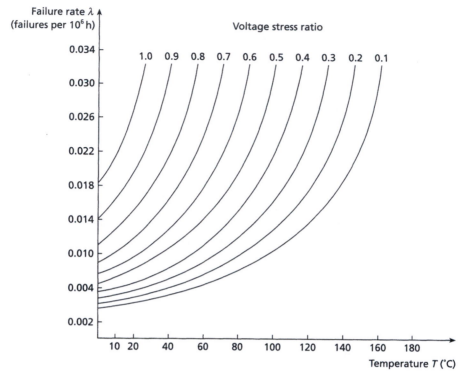

Figure 4.1 Failure rate of silicon npn transistor as a function of temperature and voltage stress ratio

failure rate increases with temperature; the rate of increase is determined by the activation energy E, the higher E the higher the rate of increase. Typical values of E vary between 0.25 and 1.5 eV depending on the failure process. One important failure process in semiconductors is where increasing temperature causes higher diffusion rates of charge carriers resulting in the wrong distribution of carriers in the P and N regions. The temperature T of the component depends on the ambient temperature of the air surrounding the component, the electrical power produced in the component, and the amount of convective heat transfer between the component and air. In general the temperature of the component will be greater than that of the surrounding air. The failure rate of any element or system which uses electronic components will depend critically on the temperature of those components.

Temperature can also affect the failure rate of other elements and components. The electrical insulation properties of materials deteriorate at both high and low temperatures. The rates of unwanted chemical reactions, of which corrosion is one example, can increase markedly as temperature increases. Mechanical components expand/contract as temperature is increased/decreased resulting in wear and damage.

4.3.3 Environment

Temperature is just one example of environmental factors which can affect the failure rate of elements and components. Other environmental factors which influence failure rate are:

Humidity of atmosphere
Salt in atmosphere
Dust in atmosphere
Exposure to frost
Nature of process material, *e.g.* corrosive, erosive, dirty, multiphase
Vibration
Mechanical shock
Thermal shock
Electromagnetic radiation

It should be noted that when two of these factors occur together the resulting failure rates may be especially high. For example, the presence of salt in an atmosphere which is also very humid will cause corrosion.

Table 4.5 illustrates the effect of environment on failure rate by comparing the failure rates of instruments which are in contact with the process fluid, with those not in contact with the process fluid, at a given chemical works.[3] It can be seen that the failure rate of those instruments in contact with the process fluid is significantly higher than those in contact with the background plant environment.

Table 4.6 compares the failure rates of two given types of instrument, *i.e.* control valves and differential pressure transmitters, when used with 'clean' and 'dirty' fluids at a given chemical works.[3] The 'clean' fluid is a single-phase, non-corrosive gas or liquid with no

Table 4.5 Failure rates of instruments in contact and not in contact with process fluids (after Lees[3])

	No. at risk	No. of faults	Failure rate (faults/year)
Instruments in contact with process fluids	2285	1252	1.15
Pressure measurement	193	89	0.97
Level measurement	316	233	1.55
Flow measurement	1733	902	1.09
Flame failure device	43	28	1.37
Instruments not in contact with process fluids	2179	317	0.31
Valve positioner	320	62	0.41
Solenoid valve	168	24	0.30
Current–pressure transducer	89	23	0.54
Controller	1083	133	0.26
Pressure switch	519	75	0.30

Table 4.6 Failure rate of two types of instrument when used with clean and dirty fluids (after Lees[3])

Instrument	No. at risk	No. of faults	Failure rate (faults/year)
Control valve			
Clean fluids	214	17	0.17
Dirty fluids	167	71	0.89
Differential transmitter			
Clean fluids	27	5	0.39
Dirty fluids	90	82	1.91

impurities present; the 'dirty' fluid has either solid particles or sludge present or may polymerize or corrode.

Table 4.7 shows the observed failure rates for given types of instruments at three different chemical works A, B and C which process different types of materials and have different background plant environments.[3] The observed failure rates can be regarded as the product of a **base failure rate** λ_B and an **environmental correction factor** π_E, *i.e.*

$$\lambda_{OBS} = \pi_E \times \lambda_B \tag{4.5}$$

Here λ_B corresponds to the best environmental conditions and π_E has values 1, 2, 3 or 4; the highest figure corresponding to the worst environment.

4.3.4 Stress

If a mechanical element, for example a beam or strut, is subject to a stress which produces a corresponding strain greater than the elastic limit of the element material, then the element will undergo plastic deformation and fail. Stated simply, if applied stress exceeds element strength then failure occurs. These concepts of 'stress' and 'strength' apply to all types of elements and components. **Stress** can refer to any variable which is applied to an element or component; for example, mechanical stress, pressure, voltage or torque. **Strength** can refer to any property of the element which resists the applied stress; for example, elastic limit, rated voltage, melting point. Other examples of stress-induced failure are:

(a) A transistor in an integrated circuit fails when the applied voltage causes a local increase in current density, which in turn causes the temperature at the location of the transistor to rise above the melting point.
(b) A hydraulic valve fails when the applied pressure exceeds the maximum pressure limit of the seal and leakage of fluid occurs.
(c) A drive shaft fractures when the applied torque produces a strain which exceeds the elastic limit.

Table 4.7 Observed failure rates for instruments in different chemical plant environments (after Lees[3])

Instrument	Observed failure rate (faults/year)	Environmental correction factor	Base failure rate (faults/year)
Control valve (p)			
—	0.25	1	0.25
Works A	0.57	2	0.29
Works B	2.27	4	0.57
Works C	0.127	2	0.064
Differential pressure transmitter (p)			
—	0.76	1	0.76
Works A (flow)	1.86	3	0.62
Works A (level)	1.71	4	0.43
Works B (flow)	2.66	4	0.67
Works C (flow)	1.22	2	0.61
Variable area flowmeter transmitter (p)			
—	0.68	1	0.68
Works A	1.01	3	0.34
Thermocouple			
Works A	0.40	3	0.13
Works B	1.34	4	0.34
Works C	1.00	4	0.25
Controller			
—	0.38	1	0.38
Works A	0.26	1	0.26
Works B	1.80	1	1.80
Works C	0.32	1	0.32
Pressure switch			
—	0.14	1	0.14
Works A	0.30	2	0.15
Works B	1.00	4	0.25

p: pneumatic

It is obvious, therefore, that to avoid failure, **stress** should be less than **strength**, the question is how much? In general stress and strength will not have fixed values but are distributed statistically. We can specify the distributions of stress x and strength y using Gaussian probability density functions $p(x)$ and $p(y)$ respectively (Section 1.2.3); these functions have corresponding means $\bar{x}$, $\bar{y}$ and standard deviations σ_x, σ_y. Figure 4.2 shows the situation when the probability density functions have equal standard deviations, *i.e.* $\sigma_x = \sigma_y = \sigma$. Figure 4.2(a) shows the situation when the separation of the means $\bar{y} - \bar{x}$ is equal to eight standard deviations, *i.e.* $\bar{y} - \bar{x} = 8\sigma$. The area of overlap between the distributions is negligible, *i.e.* the probability of stress exceeding strength and the probability of failure is negligible. Figure 4.2(b) shows the situation when the separation of the means is reduced so that $\bar{y} - \bar{x} = 6\sigma$; the area of overlap and the probability of

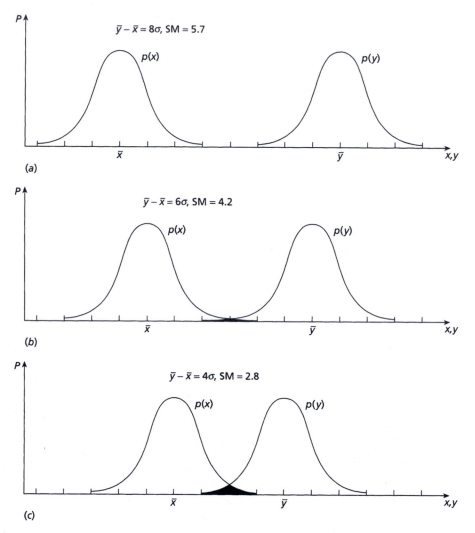

Figure 4.2 Overlapping of stress and strength distributions

failure is now significant. Figure 4.2(c) shows the situation when $\bar{y} - \bar{x} = 4\sigma$; the area of overlap and the probability of failure is now appreciable.

The separation of the distributions can be quantified using the concept of **safety margin**; this is a non-dimensional quantity defined by Carter.[4]

safety margin $$\boxed{SM = \frac{\bar{y} - \bar{x}}{\sqrt{(\sigma_x^2 + \sigma_y^2)}}}$$ (4.6)

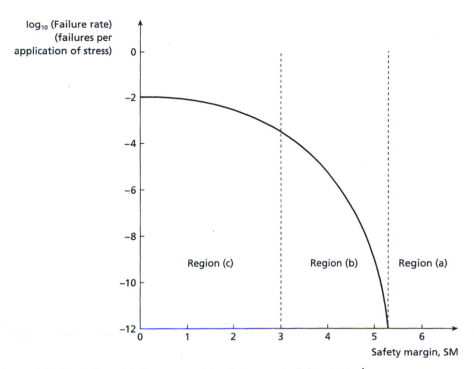

Figure 4.3 Variation of failure rate with safety margin (after Carter[4])

In the special case $\sigma_x = \sigma_y = \sigma$, we have SM $= (\bar{y} - \bar{x})/\sqrt{2}\sigma$. Thus Figure 4.2(a) corresponds to SM $= 8/\sqrt{2} = 5.7$, Figure 4.2(b) corresponds to SM $= 6/\sqrt{2} = 4.2$ and Figure 4.2(c) to $4/\sqrt{2} = 2.8$.

Figure 4.3 shows a typical variation of failure rate with safety margin, SM. Here failure rate is defined as the number of failures per application of stress and is plotted on a log scale.[4] The curve is obtained by calculating the area of overlap between the distributions in Figure 4.2 and can be divided into three regions corresponding to Figure 4.2(a), (b) and (c). In region (a), SM is greater than around 5.3 and the failure rate is virtually zero. In region (b), SM lies between approximately 5.3 and 3.0, and the failure rate increases rapidly as SM is reduced. In region (c), SM lies approximately between 3 and 0 and the failure rate is high but tends to a constant value as SM is reduced. Ideally, the safety margin should be chosen to correspond to region (a).

Figure 4.1 shows how the failure rate of a typical electronic component varies with both voltage stress and temperature. A family of curves is plotted each corresponding to different values of **stress ratio** where:

$$\textbf{stress ratio} = \frac{\text{Mean operating voltage}}{\text{Mean maximum (rated) voltage}} = \frac{\bar{x}}{\bar{y}} \tag{4.7}$$

At any operating temperature, *e.g.* 10°C, increasing the stress ratio from 0.1 to 1.0 causes a corresponding increase in λ; we note that the increase in λ for an increase in stress ratio between 0.9 and 1 is far greater than the increase in λ for an increase in stress ratio between 0.1 and 0.2. For many electronic components, the upper limit on the stress ratio is 0.7, this is referred to as **de-rating**.

4.3.5 Other factors

One factor which affects component failure rates is the **complexity and maturity of the manufacturing process**. Experimental data show that component failure rate decreases as a manufacturing process matures and component imperfections are removed.

4.4 Failure rate models

In the design of systems it is extremely useful to have **failure rate models** for each element or component which makes up the system. A failure rate model is an equation for the failure rate of the element/component which takes account of the effects of the four factors discussed above together with any other relevant factors. More progress has been made on models for electronic components than for any other devices. For simple electronic components, *e.g.* resistors or capacitors, where temperature and stress ratio are low, a simple model equation of the form:

$$\lambda = \lambda_b \pi_Q \pi_E \qquad (4.8)$$

is sufficient. Here λ_b is the base failure rate, π_Q is the component quality factor and π_E is the environment factor.

More detailed models are required for more complex electronic devices such as integrated circuits. The US military handbook MIL-HDBK-217E *Reliability Prediction of Electronic Equipment* specifies the following equation for failure rate:

$$\lambda_P = \pi_Q \pi_L [C_1 \pi_T \pi_V + (C_2 + C_3) \pi_E] \qquad (4.9)$$

where π_Q = quality factor
π_L = learning factor (based on maturity of manufacturing process)
π_T = temperature factor (based on maximum junction temperature)
π_V = voltage de-rating stress factor
π_E = environment factor
C_1, C_2, C_3 = failure rates which depend on circuit complexity
λ_P = predicted failure rate – failures per 10^6 hours

If we take the example of an LM108A, a 28 transistor linear integrated circuit with several levels of screening used in a sheltered ground environment, we have:[5]

$\pi_Q = 3, \quad \pi_L = 1, \quad \pi_T = 0.23 \text{ (at } T = 35°C)$
$\pi_V = 1.0, \quad \pi_E = 0.38 \text{ (benign ground environment)}$
$C_1 = 0.021, \quad C_2 = 0.0048, \quad C_3 = 0.0020$

giving:

$\lambda_P = 3 \times 1[0.021 \times 0.23 \times 1 + (0.0048 + 0.0020) \times 0.38]$
$= 3[0.004\ 83 + 0.002\ 58]$
$= \mathbf{0.022} \text{ failures per } 10^6 \text{ hours}$

4.5 Calculation of equipment failure rate from component failure rate data

In Section 4.2 we saw how failure rates can be found experimentally by directly measuring the frequency of failure of elements and components of a given type. If failure rate data/models for basic components are known, then the failure rate of elements or modules which are made up from these components can be calculated using Equations 3.5 and 3.8 (components in series and parallel).

Table 4.8 shows the calculation of the overall failure rate, from basic component data, for an electronic square root extractor module.[6] The module gives an output voltage signal in the range 1–5 V, proportional to the square root of the input signal with range 4–20 mA; this type of module is commonly used in fluid flow rate control systems. The module is made up from basic electronic components of various types, all connected in series. Several components of each type are present. From Equation 3.5, the failure rate λ of a module containing m different component types in series with failure rates λ_1, $\lambda_2, \ldots, \lambda_i, \ldots, \lambda_m$, and one of each type present is:

$$\lambda = \lambda_1 + \lambda_2 + \ldots + \lambda_i + \ldots + \lambda_m \qquad (4.10)$$

If there are multiple components of each type, all connected in series, then the module failure rate is given by:

$$\boxed{\lambda = N_1\lambda_1 + N_2\lambda_2 + \ldots + N_i\lambda_i + \ldots + N_m\lambda_m}$$

**module failure rate –
multiple components
of each type**   (4.11)

where $N_1, N_2, \ldots, N_i, \ldots, N_m$ are the quantities of each component type. The failure rate of each component type is calculated using the model equation:

$$\lambda_i = (F_{1i} + F_{2i} + F_{3i}) \times K_{1i} \times K_{2i} \times K_{3i} \times K_{4i} \qquad (4.12)$$

which has some similarities with Equation 4.9. Table 4.8 gives values of F_1, F_2, F_3, K_1, K_2, K_3, K_4 and failure rate λ_i for each component. Each failure rate is then multiplied by the appropriate quantity N_i and the $N_i\lambda_i$ values added together to give a total module failure rate of 3.01×10^{-6} per hour.

Table 4.9 compares calculated and observed failure rates for a range of instruments for which a large amount of component and module failure rate data has been collected.[1]

Table 4.8 Calculation of overall failure rate for electronic square root modules
(after Hellyer[6])

			Reliability analysis							
	Unit:		**M13**							
	Description:		**Square root extractor: IP4–20MA: OP1–5 V**							
Component					**Failure rates per 10^{10} hours**					
	F1	**F2**	**F3**	**K1**	**K2**	**K3**	**K4**	**λ_i**	**N_i**	**$N_i\lambda_i$**
RESISTORS										
Carbon film										
$0 < R \leqslant 100$ K	100	0	0	1	1	1	1	100	17	1700
100 K $< R \leqslant 1$ M	100	0	0	1	1.1	1	1	110	2	220
1 M $< R \leqslant 10$ M	100	0	0	1	1.6	1	1	160	3	480
Metal film										
$0 < R \leqslant 100$ K	150	0	0	1	1	1	1	150	17	2550
100 K $< R \leqslant 1$ M	150	0	0	1	1.1	1	1	165	3	495
POTENTIOMETERS										
$0 < R \leqslant 50$ K	700	0	0	1	1	1	1	700	8	5600
50 K $< A \leqslant 100$ K	700	0	0	1	1	1.1	1	770	1	770
CAPACITORS										
Metal film										
$0 < C \leqslant 33$ nF	200	0	0	1	1	1	1	200	3	600
33 nF $< C \leqslant 1$ μF	200	0	0	1	1	1.3	1	260	2	520
1 μF $< C \leqslant 10$ μF	200	0	0	1	1	1.5	1	300	1	300
Ceramic										
$0 < C \leqslant 3.3$ nF	150	0	0	1	1	1	1	150	1	150
Electrolytic										
$3.2 < C \leqslant 62$ μF	500	0	0	0.29	1	0.7	1	102	1	102
DIODES										
Silicon LP	200	0	0	0.55	1	1	1	110	2	220
Zener	1000	0	0	1.3	1	1	1	1300	1	1300
TRANSISTORS										
NPN LP	400	0	0	1.4	1	1	1	560	2	1120
INTEGRATED CIRCUITS										
OP AMP	160	50	600	1	1	1	1	810	9	7290
Quad switch	38	320	560	1	1	1	1	918	1	918
OTHERS										
Edge connectors	300	0	0	1	1	1	1	300	8	2400
Soldered joints	20	0	0	1	1	1	1	20	167	3340
PCB	60	0	0	1	1	1	1	60	1	60
		Total λ:	3.01 × 10^{-6} per hour							
		MTBF:	3.32 × 10^5 hours							

Notes
(a) Data sources:
  1: *Electronic Reliability Data*, National Centre of Systems Reliability, Application Code 2, 25C
  2: *Reliability Prediction Manual for Guided Weapon Systems*, MOD
  3: Component supplier's information
(b) Rate = (F1 + F2 + F3) × K1 × K2 × K3 × K4

Table 4.9　Comparison of calculated and observed failure rates for a range of instruments (after Wright[1])

Instrument	Failure/year		Ratio of observed to predicted
	Predicted	Observed	
Pneumatic			
Flow transmitter	0.58	0.22	0.38
Level transmitter	0.63	0.35	0.56
Pressure transmitter	0.45	0.76	1.7
Differential pressure transmitter	0.59	0.94	1.6
Pressure gauge	0.088	0.032	0.36
Electronic			
Oxygen analyser type 1	3.2	4.5	1.4
Oxygen analyser type 2	1.9	1.2	0.63
Temperature trip amplifier	2.8	2.3	0.82
Pulse rate measuring channel	17.8	17.4	0.98
Gamma monitor	0.38	0.34	0.89
Computers			
Small computer type 1 + tape reader	27.5	23	0.84
Small computer type 2	9.75	16.6	1.7

The agreement is good for purely electronic units, such as the temperature trip amplifier and pulse rate measuring unit. It is much worse for the transmitters which are in contact with the process fluid and therefore more difficult to quantify the environmental factors exactly.

4.6　Software reliability

4.6.1　Introduction

Computers are now widely used in all branches of engineering. Many industrial processes, *e.g.* steel, chemicals, food, oil, gas and electricity generation, rely on computers to monitor and control critical process variables. The reliability of this control is therefore dependent on the reliability of the computer. Furthermore microcomputers form an integral part of a wide range of electronic systems. These 'embedded' microcomputers use computer programs, *i.e.* 'software', to perform functions previously performed by electronic circuits, *i.e.* 'hardware'. For example, the calculation of a square root can be implemented using either hardware, *e.g.* the complex electronic circuit of Table 4.8, or software, *e.g.* a single statement in a high level programming language. Performing functions in software rather than hardware can therefore lead to a simpler, more robust overall system.

Since software is vital to the performance of a large number of engineering functions, its reliability should be closely studied. Each copy of a computer program is identical to the original, so that failures due to product variability, often common with hardware, cannot occur with software. Also, unlike hardware, software cannot usually degrade with time; in the few special cases that degradation does occur it is easy to restore to the original standard. However, software can fail to perform the required function due to undetected errors in the program. If a software error exists in the original program then the same error exists in all copies. If this error produces a failure in a certain set of circumstances, then failure will always occur under these circumstances with possibly serious consequences. Many programs consist of a large number of individual statements and logical paths so that the probability of a significant number of errors being present is high. A software reliability effort is therefore required to minimize the number of errors present by the use of systematic programming methods, checking and testing.

4.6.2 Software reliability model

In spite of all efforts to ensure that the software is free from errors, some residual errors ('bugs') often persist. Shooman has produced a simple model for software reliability which assumes that the average rate at which software bugs are detected and corrected from similar programs is approximately constant.[7] The software failure rate will then be proportional to the number of remaining bugs. Thus if we assume: (a) no new bugs are created during the debugging process and (b) all detected errors are corrected then we have:

Fractional number of residual bugs
= Fractional number of total bugs − Fractional number of corrected bugs

i.e.

$$\varepsilon_R(\tau) = \left(\frac{E_T}{I_T}\right) - \varepsilon_C(\tau) \tag{4.13}$$

where τ = debugging time in months
  E_T = total number of errors
  I_T = total number of instructions

Shooman's experimental findings suggest that the ratio (E_T/I_T) is approximately constant and lies in the range

$$3 \times 10^{-2} \le \left(\frac{E_T}{I_T}\right) \le 1.0 \tag{4.14}$$

In his model the fractional number of corrected bugs $\varepsilon_C(\tau)$ is proportional to τ, *i.e.*

$$\varepsilon_C(\tau) = \rho\tau \tag{4.15}$$

where ρ = fractional rate at which errors are removed per month. Again ρ is approximately constant and lies in the range:

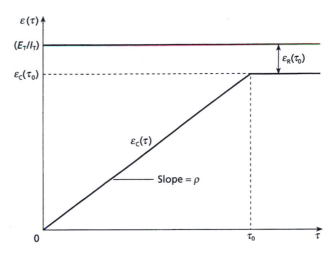

Figure 4.4 Shooman model for software reliability

$$3 \times 10^{-3} \leqslant \rho \leqslant 1.0 \text{ per month} \tag{4.16}$$

If the debugging process is concluded at time τ_0, $\varepsilon_C(\tau)$ will therefore remain at the constant value $\varepsilon_C(\tau_0)$ and $\varepsilon_R(\tau_0)$ will correspondingly remain at the constant value (Figure 4.4):

$$\varepsilon_R(\tau_0) = \left(\frac{E_T}{I_T} \right) - \varepsilon_C(\tau_0) \tag{4.17}$$

The failure rate λ of the software is then proportional to $\varepsilon_R(\tau_0)$, the fractional number of bugs left in the program, *i.e.*

$$\lambda = K\varepsilon_R(\tau_0) \tag{4.18}$$

The values of (E_T/I_T), ρ and K must be found by experimental testing before a value of λ can be established. Shooman's model is one of many attempts to quantify the reliability of software.

4.6.3 Methods of improving software reliability

Specification

Many of the errors recorded during software development originate in the specification. The software specification must describe fully and accurately the requirements of the program. There are no safety margins in software design as in hardware design. The specification must be logically complete, *i.e.* must cover all possible input conditions and output requirements.

As an example of what can go wrong if the specification is incomplete, consider a program to find the middle value, *i.e.* the value that is neither the highest nor lowest, of three transducer signals X, Y, Z (Example 3.1). Figure 4.5 is a flowchart showing the logic. If we have $X = 11.7$ mA, $Y = 11.3$ mA and $Z = 11.9$ mA, *i.e.* $Z > X > Y$, then X is selected. If the X transducer fails, so that $X = 0$ mA, $Y = 11.3$ mA and $Z = 11.9$ mA, *i.e.* $Z > Y > X$, then Y is selected. The system therefore survives the failure of one transducer and is more reliable than the corresponding single channel system. If both X and Y transducers fail, *i.e.* $X = 0$ mA, $Y = 0$ mA and $Z = 11.9$ mA, the program should select Z: however, since this possibility has not been considered in the specification, it is uncertain which value will be selected. An extra logical operation is required which selects the non-zero signal if the other two signals are zero.

Software system design

This follows from the specification and is often a flow chart (Figure 4.5) which defines the program structure, test points, limits, *etc.* To be reliable a program must be *robust*, *i.e.* it should be able to survive error conditions without serious effect such as 'crashing' or becoming locked in a loop.

Structure

Structured programming is a methodology that forces the programmer to use certain clear, well-defined methods of program design, rather than allow complete freedom to design intricate, complex programs which are prone to error and difficult to understand and debug. A major source of error is the use of the GOTO statements in loops and branches. At several points in Figure 4.5 two numbers, *e.g.* X and Y, are compared, if X is greater than Y the program branches one way, if Y is greater than X the program branches another way. Figure 4.6 shows how the above branch instruction can be programmed (using BASIC) in either an unstructured or a structured way. The unstructured method is sensitive to errors in the line numbers and it may be difficult to trace the line numbers back to the decision point. The structured method eliminates this dependency on line numbers and is easier to understand and check.

Modularity

Modular programming breaks a program down into smaller, individual blocks or modules each one of which can be separately specified, written and tested. This makes the program much easier to understand and check. The middle value selector program of Figure 4.5 is an example of such a module.

Fault tolerance

Programs should be written so that if an error does occur, the program should be able to find its way out of the error condition and indicate the source of the error. This can

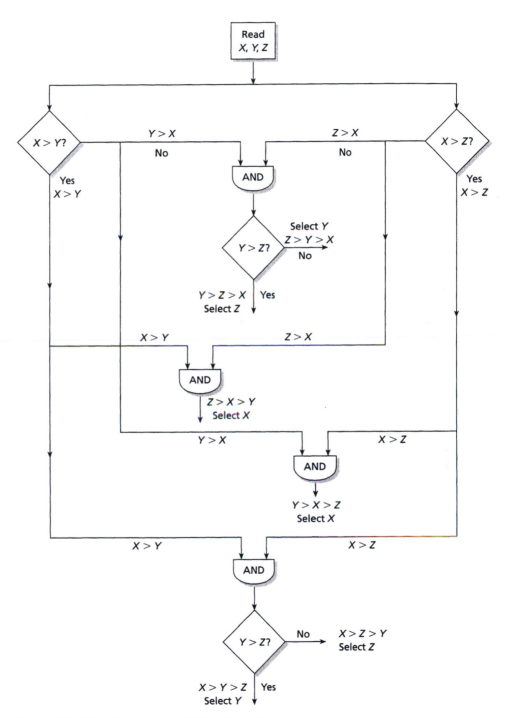

Figure 4.5   Flow diagram for middle value selector program

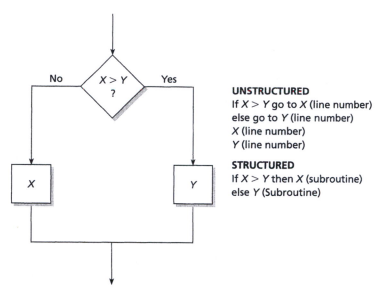

Figure 4.6 Structured and unstructured programming

be done, for example, by checking each program module in turn by comparing the value of the variable at the module output with a specified range of values: if a variable is out of range then the module which contains the error can quickly be identified. In applications where safety is vital, for example where a computer is used to control an industrial process, if an error does occur the program should first detect it, send an alarm message and then move the process to known safe conditions. Fault tolerance can also be provided using **program redundancy**. Here separately coded programs, *i.e.* programs written to the same specification by different programmers, are either run simultaneously on separate computers or at different times on one computer. A voting or selection procedure then decides which output is to be used (Section 3.4 and Figure 4.5). This method is based on the assumption that it is extremely unlikely that separately coded programs will contain the same coding errors but provides no protection against errors in the program specification.

Languages

The reliability of software also depends on the computer language used. There are two main types of language in industrial use:

(a) Assembly-level programming
(b) High-level language programming

Assembly-level programs are faster to run and require less memory than high-level programs and may be preferred for real-time systems. However, assembly-level programming is far more difficult; a large number of detailed instructions, which are specific to a given processor, are required in order to perform a given operation. This makes checking far more difficult; also, there are several types of error in assembly-level programming which cannot be made in high-level language.

High-level languages, for example, Fortran, BASIC, Ada, Pascal, are processor-independent and operate using a **compiler** which translates the high-level language into the assembly language of a given processor. High-level languages therefore run more slowly and require more memory; programming and checking are, however, much simpler. The older high-level languages, *e.g.* Fortran and BASIC, do not encourage structured programming. BASIC, for example, allows GOTO statements which are very error prone. The newer languages, *e.g.* Pascal and Ada, strongly encourage structured programming and discourage the use of GOTO statements; less error-prone statements such as IF . . . THEN . . . ELSE are used as an alternative.

Software checking

Modern high-level language compilers have error detection capability so that errors of logic, syntax and other coding errors can be detected and corrected before an attempt is made to load the program. When the program is capable of being run, it is then necessary to confirm that it meets all requirements of the specification under all anticipated input conditions.

4.7 Human reliability

The subject of human reliability is a wide one; a large amount of data has been collected on the reliability of human response in a number of different situations. However, no adequate models exist which will enable the failure rate of a given human, carrying out a given operation, to be accurately predicted. This is because human failure rates for a given operation depend on a large number of factors which can be conveniently grouped under three main headings: **intrinsic**, **environmental** and **stress**.

Intrinsic factors

These cover the basic characteristics of an individual and include:

Motivation – *i.e.* does the person want to perform the operation correctly?
Physical ability – *i.e.* is the person physically capable of performing the operation?
Mental ability – *i.e.* has the person the basic intelligence required to perform the operation?
Temperament – *i.e.* can the person remain sufficiently calm under stress to perform the operation?
Concentration – *i.e.* can the person exclude all other influences while performing the operation?
Speed of response – can the person respond quickly enough in an emergency situation?
Knowledge – has the person sufficient knowledge to carry out the operation correctly?

Ideally, the characteristics required for a given operation or job should first be clearly identified and the selection process then designed to choose people with these characteristics. The selected personnel should then undergo a course of training, which gives them the technical knowledge and relevant experience necessary for performance of the operation or job.

Environmental factors

The reliability of a given individual, performing a given operation, will depend on the **total environment** in which the individual is working. This total environment is made up of **physical**, **organizational** and **personal** factors.

Physical factors include: temperature, humidity, noise level, dirt, time of day (*e.g.* night shift). **Organizational** factors include: relationship with colleagues, relationship with supervisor/management, job satisfaction, salary/wages, job security, promotion prospects. **Personal** factors include hunger, thirst, tiredness, physical and mental health, home life.

Stress factors

The reliability of a given person performing a given task in a given environment depends on their **stress level**. Figure 4.7 shows qualitatively the relation between human error rate (expressed in number of errors per operation) and stress level. It can be seen that there is an optimum stress level at which the error rate is minimum; if the person is either bored or overexcited the error rate increases. A recent UK aircraft crash shows the effect of stress on error rate; in a situation where one engine was on fire and one was working normally, the crew shut down the good engine.

Research also indicates that human error rate increases significantly as the complexity of the operation or task increases. Some suggested error rates are:

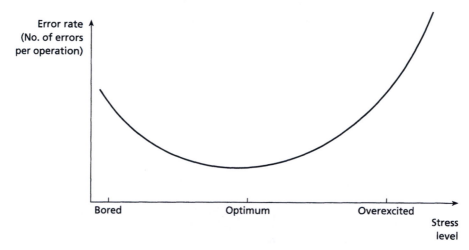

Figure 4.7   Effect of stress level on human error rate

10^{-4} to 10^{-3} for simplest possible tasks, *e.g.* failure to respond to an annunciator, reading a single alphanumeric digit wrongly.

10^{-3} to 10^{-2} for routine simple tasks, *e.g.* failure to read analogue indicator correctly, failure to correctly replace a printed circuit board.

10^{-2} to 10^{-1} for routine tasks requiring care, *e.g.* typing in a character wrongly, failure to reset a valve after a related task.

10^{-1} to 1   for complicated non-routine tasks, *e.g.* failure to notice position of valves, failure to act correctly after one minute in an emergency situation.

Summary

In general there is a large amount of good failure rate data for most engineering components and elements. This is especially so for electronic components where detailed failure rate models exist. However, it is far more difficult to predict software and human reliability.

References

1. Wright, R. I. (SRD Warrington) (1984) Instrument reliability, *Instrument Science and Technology*, Vol 1, Institute of Physics, Bristol, pp. 82–92.
2. Smith, D. J. (1988) *Reliability and Maintainability in Perspective*, 3rd edn, Macmillan, Basingstoke, pp. 243–9.
3. Lees, F. P. (1976) The reliability of instrumentation, *Chemistry and Industry*, March, pp. 195–205.
4. Carter, A. D. S. (1986) *Mechanical Reliability*, 2nd edn, Macmillan, Basingstoke, pp. 20–63.
5. National Semiconductor Corporation (1987) *The Reliability Handbook*, 3rd edn, California, Santa Clara, January, pp. 131–2.
6. Hellyer, F. G. (Protech Instruments and Systems) (1985) The application of reliability engineering to high integrity plant control systems, *Measurement and Control*, **18**, June, pp. 172–6.
7. Shooman, M. L. (1979) In T. Anderson and B. Randell (eds) *Computing Systems Reliability*, CUP, Cambridge.

Self-assessment questions

4.1 A CMOS integrated circuit has a failure rate of 0.1 per 10^6 hours at 20° C. If the activation energy for the failure process is 0.5 eV, find the failure rate at 50° C.

4.2 The elastic stress limits for a given type of mechanical component have a Gaussian probability density function with a mean of 6 M Pa and a standard deviation of 1 M Pa. The components are subject to applied stresses described by a Gaussian probability density function with a mean of 4 M Pa and a standard deviation of 2 M Pa.

(a) Calculate the safety margin.

(b) Use Figure 4.3 to estimate the failure rate.

4.3 From a reliability point of view, a simple pressure transducer can be regarded as consisting of one of each of the following components in series:

Bellows
Fulcrum
Rectangular shaft
Helical spring
Seal
Transformer (small signal)
Voltage regulator

Use the data given in Tables 4.1 and 4.2 to estimate the annual failure rate for the transducer.

4.4 Figure Q4.4 shows an 8-bit digital-to-analogue converter. The components present and their failure rates are as follows:

Component	Failure rate λ (per 10^{10} hours)
Resistor – carbon film (< 100 kΩ)	100
Resistor – carbon film (> 100 kΩ)	110
Op-amp integrated circuit	800
Transistor switch S	560
Voltage reference	2000
Soldered joint –•–	20

Calculate the annual failure rate for the converter.

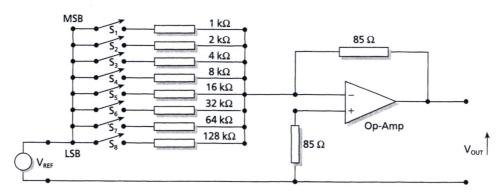

Figure Q4.4

4.5 A computer program has 2000 instructions and contains 137 errors. Errors are removed at the fractional rate of 0.01 per month. The constant of proportionality between annual failure rate and fractional number of errors left in the program is 0.05. If the debugging process is concluded after five months, estimate the resulting failure rate of the program.

5

Quality and reliability in design and manufacture

5.1 | Introduction

In Chapter 1 we saw how the quality of an engineering product could be measured using a set of continuous performance characteristics $\{x\}$. These characteristics then lead to a product specification in terms of a set of target values $\{x_T\}$ and sets of specification limits $\{LSL\}$, $\{USL\}$. In Chapter 2 we saw that reliability can be specified by mean failure rate $\bar{\lambda}$. This chapter begins by reviewing the **methods** and **culture** necessary to achieve quality. It then goes on to discuss the factors that determine profit and how these factors in turn are influenced by quality and reliability. It then shows how the essential engineering activities of **design**, **manufacture** and **testing** can lead to improved quality and reliability.

5.2 | The achievement of quality

5.2.1 Total quality management

Figure 5.1 shows that in order to achieve **Total Quality Management (TQM)** an organization must have both the **methods and culture** necessary for quality. The methods involve a set of systems and procedures. The culture of an organization is the set of values, behaviour and standards which make it tick and therefore, as shown in Figure 5.1, involves both **management and workforce**. A quality culture can only develop if management are committed to it; *i.e.* are willing to install an appropriate quality management structure, are willing to accept responsibility for quality and are willing to provide the necessary resources. Similarly a quality culture can only develop if the workforce

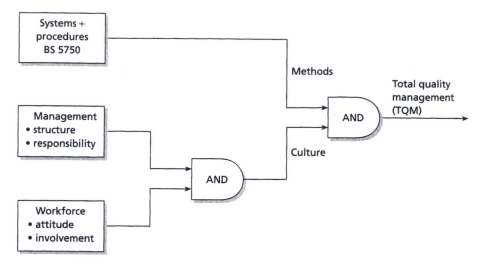

Figure 5.1   The achievement of total quality management (TQM)

are committed to it; *e.g.* are willing to become involved in quality, are willing to accept changes in working methods and are willing to adopt appropriate attitudes. These three essential elements of methods, management and workforce are now discussed in more detail.

5.2.2 Quality methods

Figure 5.2 shows a typical **life cycle** for an engineering product showing the different states from initial specification through design, production, distribution, installation, operation, and maintenance to final disposal after use. BS 5750 (ISO 9001) uses this life cycle to define a **quality system;**[1] this is a set of procedures which are applicable to each of the stages and aim to improve the quality of the final product. The following are some important examples of these procedures:

(a) Production of a clearly defined specification which is a statement of the quality objectives for the product.

(b) Contractual agreements with customers as to how their requirements will be met.

(c) Development of a 'robust' design which meets the specification.

(d) Testing of the design to ensure that the specification is met.

(e) Systematic design review with control of design changes and test of the redesign against the specification.

(f) Verification that all documents, drawings, specifications *etc.* are up to date and correct.

(g) Specification of components and materials to be procured.

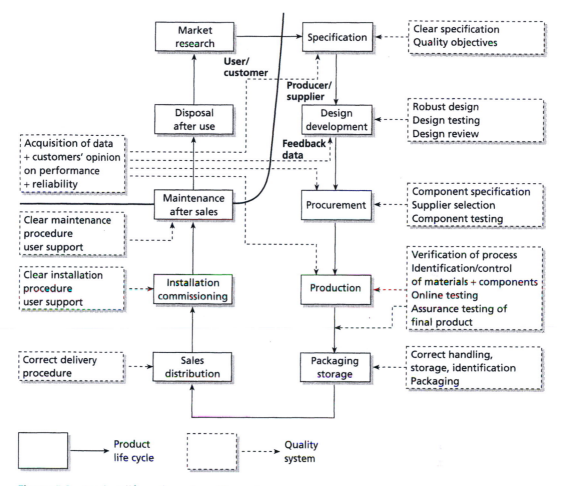

Figure 5.2 Product life cycle and quality system

(h) Selection of suppliers of components and materials.

(i) Verification that incoming materials and components meet specification and are documented.

(j) Control of 'free issue' material.

(k) Verification that the production process is capable of meeting the requirements of the design.

(l) Identification of materials, components and products as they pass through the process.

(m) Methods to ensure that manufacturing equipment is set up and maintained correctly to perform the desired function.

(n) Control of the process to ensure that the product specification is met.

(o)   Inspection and testing of the product as it proceeds through the process.
(p)   Testing of the completed product to verify that it meets the specification.
(q)   Verification that measurement, test and inspection equipment meets its specification and is calibrated against standards traceable back to National Standards.
(r)   Methods to ensure that all materials, components and products not conforming to specification are clearly identified and segregated.
(s)   Full analysis of the root cause of faults and correction to prevent any recurrence.
(t)   Methods to ensure the correct handling, storage, identification, packaging, delivery and installation of the product.
(u)   Provision of after-sales technical support for installation, commissioning and maintenance.
(v)   Acquisition of data on the operating performance and reliability, *etc.*, of the product once installed and commissioned.
(w)   Acquisition of customer opinion and market research information.
(x)   Use of all feedback data in the development of the specification for a replacement product.
(y)   Use of appropriate statistics at different points in the product life cycle.

5.2.3 Management for quality

The first requirement is for the appointment of a Quality Manager with a clearly defined responsibility for all aspects of quality. The Quality Manager should respond directly to the General Manager/Managing Director and should ideally be at the same level as Design, Production and Sales Managers, *etc.* (Figure 5.3). The Quality Manager should certainly not report to the manager of any department where monitoring of quality is required. The responsibilities of the Quality Manager must include:

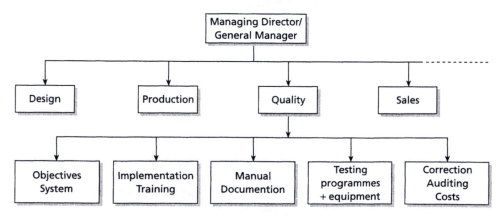

Figure 5.3   Management structure and responsibility

(a) The setting of **quality objectives**, *i.e.* the production of a specification for each of the company's products.
(b) The **formulation of a quality system** to achieve these objectives, this to be based on the procedures and methods specified in BS 5750 and summarized above.
(c) The **implementation** of that quality system.
(d) The production of a **quality manual** which defines the quality system.
(e) The **education and training** of all personnel so that they both fully understand the methods and procedures of the system and are technically competent to carry them out.
(f) The **documentation** of methods and procedures; this will include inspection reports, test reports and data, audit reports, material review reports, calibration data, quality cost reports.
(g) Maintenance of adequate **records** covering all transactions affecting the delivered product, *e.g.* input materials and components, manufacturing history, delivery documents.
(h) **Testing programmes** and calibration/maintenance of **test, measurement and inspection equipment**.
(i) The taking of **corrective action** in all situations which could affect the quality of the delivered product.
(j) The **auditing** of the quality system to check that it is operating as specified: this should be both **internal**, *i.e.* a self-audit procedure within the quality system itself; and **external**, *i.e.* by an outside, independent organization or individual.
(k) Identifying and monitoring the **costs and financial benefits** associated with the quality system.

5.2.4 Workforce commitment to quality

As stated above, management has first the responsibility to ensure that the workforce both clearly understands the objectives and methods of the quality system and is technically capable of implementing them. The next more difficult problem is to get the workforce to accept full responsibility for the quality of their products. This will involve the motivation to set and achieve higher quality standards, the flexibility to accept new working methods and procedures and the ability both to be self-critical and to accept criticism from others.

Inspection was one of the earliest methods of controlling the variation in the products of human manufacture (*i.e.* minimizing the standard deviation vector $\{\sigma\}$). Here the inspector is independent of the manufacturing effort and has the authority to reject product items which are outside specification. The entire reliance on human inspection for quality control has, however, three major disadvantages:

(a) The inspection is not perfect; the inspector is only human, and can fail to detect defects in complex situations, *e.g.* a large number of soldered joints.

(b) If the inspector has the entire responsibility for quality control then production workers may abdicate this responsibility. They can concentrate on maximizing quantity and therefore earnings.

(c) Since inspectors are paid more than normal production workers, inspection can be expensive.

A potentially more successful method is **operator control**. Here production workers are responsible for monitoring the quality of the products of the operations they perform and taking corrective action on the operations as necessary. For example, an operator measures the dimensions of a metal product from a milling machine and checks they conform to specification; if the specification is not met the machine is re-adjusted. **Statistical process control charts** (Section 1.3) are useful here. In computer controlled machine tools these charts are produced and displayed automatically and give the operator a clear indication of the state of the operations he or she controls.

Another policy is for the company to introduce a **zero defects** target. Workers are exhorted to strive for this target using a publicity campaign with prizes, posters *etc*. This evangelical approach may not be appropriate for the workforce of every company.

A fourth method is the **quality circle**. Here production workers in similar or related activities are organized into small voluntary groups. These are trained to monitor quality performance, discuss problems/solutions and make recommendations to management. These recommendations must be thoroughly assessed and implemented if technically and economically feasible.

5.3   Economics of quality and reliability

5.3.1   Profit from manufacture

As shown in Section 5.2.2, the first stage in the life cycle of an engineering product is to draw up a clearly defined specification for the product. This will be based on the company's experience with other products, market research information, customer experience and requirements. The **product quality specification** can be defined by a set of target values $\{x_T\}$ and sets of specification limits $\{LSL\}$, $\{USL\}$ (Section 1.2.2). Similarly the **product reliability specification** is defined by a target failure rate λ_T.

Figure 5.4 shows a simplified model of the factors that determine the profit associated with the manufacture of a given product and the complex interactions between them. The total manufacturing cost C_T depends critically on the above quality and reliability specifications. Because of the need to maintain an adequate profit margin on each item sold, C_T has a major effect on the selling price per item s: s is also influenced by market conditions. The number of units sold N will depend on customer satisfaction with the product, *i.e.* the product specifications, the unit selling price s and market conditions. The total sales income S_T is then given by:

$$S_T = sN \tag{5.1}$$

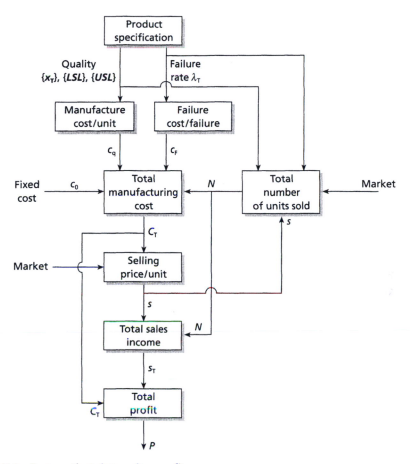

Figure 5.4 Factors that determine profit

and the profit P by:

$$P = S_T - C_T \tag{5.2}$$

The following sections discuss these factors in more detail and the influence of quality and reliability on them.

5.3.2 Total manufacturing cost C_T

The objective of the manufacturing operation can be defined as the manufacture of a product which meets the specification $\{x_T\}$, $\{LSL\}$, $\{USL\}$, $\{\lambda_T\}$ at minimum total cost C_T.

C_T consists of three cost components:

Table 5.1   Elements of manufacturing cost

Product life cycle	Quality system	
Specification	Quality manual	(P)
	Quality/reliability training	(P)
Design	Design testing – robust design	(A)
Development	Design review	(P)
	Design changes	(F)
Procurement	Component supplier selection	(P)
	Component testing	(A)
Production	Material/component identification	(P)
	Online testing	(A)
	Rework/rectification	(F)
	Scrap, defects	(F)
Packing + Storage	Final product testing	(A)
	Test equipment maintenance	(A)
Sales + Distribution	Reinspection/re-examination	(F)
	Audits	(P)
Installation	Installation testing	(A)
Commissioning	Commissioning failure	(F)
	Customer rejects	(F)
	Warranty claims	(F)

P = prevention
A = appraisal
F = failure

(a)   A fixed cost c_0 which is independent of the number of units N produced. This could incude, for example, the costs of plant, machinery and labour.

(b)   A variable cost which is proportional to the number of units N produced. Thus if c_Q is the manufacturing cost per unit, then the variable cost is $c_Q N$. This unit cost will depend on the quality specification; the higher the quality, the higher c_Q.

(c)   The cost of failures. This component will depend on the reliability specification, *i.e.* will be proportional to target failure rate λ_T. For N units the number of failures per year is $\lambda_T N$; if c_F is the cost incurred for each failure then the cost of failures is $c_F \lambda_T N$. Cost c_F will include, for example, the cost of replacement under warranty.

Therefore total manufacturing cost is given by:

$$C_T = c_0 + c_Q N + c_F \lambda_T N \tag{5.3}$$

Table 5.1 shows major elements of manufacturing cost; each of these will contribute to the above fixed, variable and failure cost components.

The elements are grouped under two main headings, those associated with the product life cycle and those associated with the quality system. The product life-cycle cost comprises costs for all the major activities in manufacture, *e.g.* design, development, production and sales, and accounts for between approximately 85% and 95% of the total cost C. It depends critically on the product specification; the design, procurement and production costs associated with the manufacture of a high performance sports car are much higher than the corresponding costs for a family saloon. The quality system costs are those incurred in ensuring that the delivered product meets the specification and account for approximately between 5% and 15% of C, with the average being around 8%.[2]

Table 5.1 shows that quality system cost components can be further separated into three types:

Preventive (P) – Costs of preventing failures
Appraisal (A) – Costs of testing, inspection and measurement
Failure (F)   – Costs of failure – redesign, rework, scrap

The aim here is that increased investment in prevention activities will be repaid several times over in reduced failure costs, thus resulting in an overall reduction in total quality system costs.

5.3.3 Selling price per unit *s*

The selling price per unit is determined by the total manufacturing cost C_T and market conditions. Obviously s needs to exceed C_T/N, the manufacturing cost per unit, by an adequate profit margin; the size of this margin will, however, depend on market conditions.

In a highly competitive market, where customers have a wide, free choice with plenty of price information, s will be virtually fixed, any increase in s resulting in a corresponding reduction in number sold, N. In a new, buoyant market it may be possible to increase s considerably before there is any reduction in N. Under depressed market conditions it will be necessary to reduce s and the profit margin in order to maintain N. Because of this interaction between s and N, s should be chosen to maximize profit $P = Ns - C_T$ under the prevailing market conditions.

5.3.4 Number of units sold *N*

As stated above, N will depend on both selling price s and market conditions; obviously N will be high if s is low and the market is buoyant. However, N will be largely determined by customer/user satisfaction with the product, *i.e.* the quality and reliability of the product, which is defined by the specification $\{x_T\}$, $\{LSL\}$, $\{USL\}$, λ_T. It is important to have a quantitative measurement of customer satisfaction. This may not be possible in some situations; for example, it may only be possible to judge the visual appeal of a product using the five numbers 5 (excellent), 4 (good), 3 (satisfactory),

2 (unsatisfactory), 1 (repulsive). This assessment of customer satisfaction is therefore both **discrete** (or qualitative) and **subjective**, *i.e.* dependent on the opinion of an individual customer. However, where the quality of a product is defined by a set of continuous variables $\{x\}$, customer/user satisfaction will be maximum when $\{x\}$ is equal to the target vector $\{x_T\}$. This satisfaction will reduce as $\{x\}$ deviates randomly from $\{x_T\}$ due to random variations in both individual product items and environmental conditions (Section 1.2.2). This level of satisfaction can then be quantified using the costs incurred by the customer/user.

Example 5.1

A manufacturer markets six different types of current transmitter at quality levels 1 to 6. Table 5.2 gives the following data for each quality level:

- (a) Variable manufacturing cost per unit c_Q £
- (b) Target failure rate λ_T per year
- (c) Manufacturer's cost per failure c_F £
- (d) Fixed manufacturing cost c_0 £
- (e) Number of units manufactured per year N
- (f) Selling price per unit s £
- (i) Complete the table by calculating:
  - (g) Total manufacturing cost C_T £
  - (h) Total sales income S_T £
  - (i) Manufacturer's profit P £
- (ii) At what quality level is the manufacturer's profit maximum?

Solutions

(i)

Table 5.2 Data for Example 5.1.

Quality	c_Q £	λ_T Yr^{-1}	c_F £	c_0 £1000	N 1000s	C_T £1000	s £	S_T £1000	P £1000
1	200	0.5	100	600	10	3100	320	3200	100
2	300	0.2	150	600	8	3240	450	3600	360
3	400	0.1	200	600	5	2700	700	3500	800
4	600	0.05	300	600	3	2445	1100	3300	855
5	800	0.02	400	600	2	2216	1400	2800	584
6	1000	0.01	500	600	1	1605	2000	2000	395

(ii) Manufacturer's profit is maximum at quality level 4; see Figure 5.5.

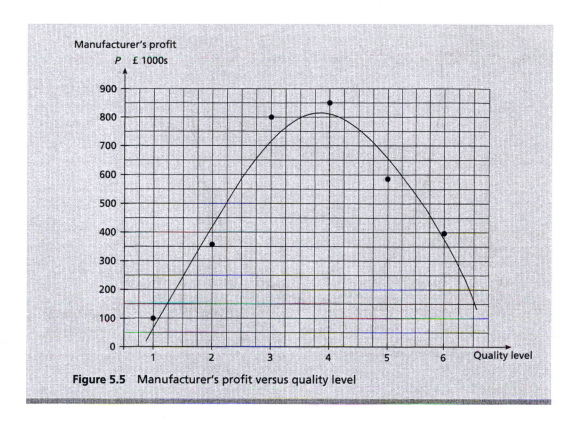

Figure 5.5 Manufacturer's profit versus quality level

5.3.5 Customer/user costs

As stated in Section 5.3.4 the level of satisfaction of a customer or user with a product can be quantified using the costs incurred by the customer/user. The most relevant measure of user costs is **total lifetime operating cost** (TLOC) or **life-cycle cost**. This is the total cost penalty incurred by the user during the lifetime of the product and is made up of three main components, *i.e.*:

TLOC = Initial cost (purchase, delivery, installation, commissioning)
    + Running costs over lifetime of product (fuel, energy, services)
    + Cost of failures and maintenance over lifetime of product

The user's satisfaction will therefore be greatest when TLOC is minimum and he will obviously examine all three cost components. The manufacturer's selling price s will have a major influence on the initial cost. Once the user has negotiated a minimum s and the product is delivered, installed and commissioned, he can concentrate on running costs and the cost of failures.

The user's running costs are determined by the deviation in $\{x\}$ from target value $\{x_T\}$. Thus, if the user buys an individual car with an urban fuel consumption of 36 m.p.g.,

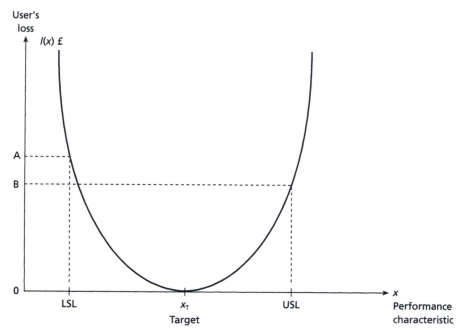

Figure 5.6   User's loss function

which is below the model target figure of 40 m.p.g., then higher petrol costs are faced. If an individual car with an above target fuel consumption of 44 m.p.g. is bought, then petrol costs are reduced. This improved fuel consumption will, however, correspond to a worsening of other characteristics such as acceleration, ease of starting and running when cold, which will again result in higher costs for the user.

We can therefore define a **user's loss function** $l(x)$. If x is the value of a continuous performance characteristic with target value x_T, then $l(x)$ represents the loss in pounds to a user caused by the deviation of x from x_T (Figure 5.6).

Therefore $l(x)$ will have a minimum value when $x = x_T$; we can define this minimum value to be zero. The loss increases as x falls below x_T, when $x = $ LSL (lower specification limit) $l = A$. Similarly l increases as x increases above x_T, when $x = $ USL (upper specification limit) $l = B$. It is important to note that with this form of loss function, the user's satisfaction is maximum when $x = x_T$ and falls off as x takes any other value, even though the value may be within the specification limits.

It is often difficult to establish the exact form of the $l(x)$ but a quadratic loss function of the form:[3]

$$l(x) = c(x - x_T)^2$$

(5.4)

may be adequate in some situations. Here c is a constant and:

$$c = \frac{A}{(LSL - x_T)^2} = \frac{B}{(USL - x_T)^2} \qquad (5.5)$$

A full analysis of the user cost of failures and maintenance over the lifetime of the product is carried out in Section 6.2. This analysis shows that, in the case of a breakdown maintenance strategy, the lifetime cost of failures is proportional to the product failure rate λ (Equation 6.2). This cost is also proportional to the mean down time, MDT, but MDT is mainly dependent on the efficiency of the user's maintenance effort and is therefore often outside the manufacturer's control. This means that if the actual failure rate of an individual item of a given product is above the target value λ_T, then the user's costs are higher than expected and their satisfaction reduced; if λ is less than λ_T, the user's costs are less than expected and their satisfaction is increased.

5.4 | Design for quality

5.4.1 Quality model

Figure 5.7 shows a **quality model** of an engineering product. The set of quantities x_1, $x_2, \ldots$, *i.e.* the vector $\{x\}$ represents the **continuous performance characteristics** of

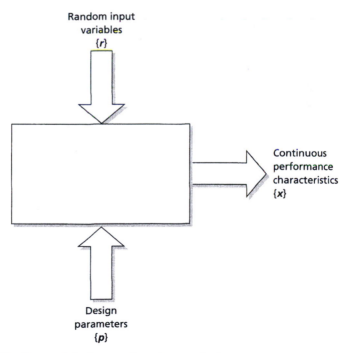

Figure 5.7   Quality model of an engineering product

Table 5.3 Environmental variables

Climatic	Electromagnetic	Mechanical	Chemical/atmospheric
Temperature	Electrical fields	Shock	Corrosive
Pressure	Magnetic fields	Acceleration	acids
Humidity	Electromagnetic radiation	Vibration	alkalis
Windspeed	infrared	single frequency	salts
Windchill	optical	several frequencies	Dust
Sunlight	ultraviolet		Biological
	microwave		fungi
	X-rays		insects
			Flammable
			Radioactive
			alpha
			beta
			gamma

the product. These have already been explained in Section 1.2.1, and Table 1.1, for example, gives continuous performance characteristics for a family car.

The set of quantities $r_1, r_2, \ldots$, i.e. the vector $\{r\}$ represents those **random input variables** which have a effect on $\{x\}$. In many cases these random inputs are caused by the environment in which the product is placed. Table 5.3 is a list of some environmental variables which could randomly affect the performance of a product.

The set of quantities $p_1, p_2, \ldots$, i.e. the vector $\{p\}$ represents the **design parameters** for the product. Examples of design parameters are the values of resistance, capacitance and inductance in an electrical circuit; values of mass, spring stiffness and damping constant in a mechanical system.

The basis of the model is that the continuous performance characteristics $\{x\}$ depend on both design parameters $\{p\}$ and random input variables $\{r\}$, i.e.

$$\{x\} = f[\{p\}, \{r\}] \tag{5.6}$$

Thus the urban fuel consumption of a family car in m.p.g. will depend on design parameters, such as the cubic capacity of the engine, weight of the car and drag coefficient, and random variables, such as wind speed, wind direction, quality of the petrol, topography of road and acceleration/braking pattern.

This type of quality model can also be used to describe manufacturing **processes**. Thus in the baking of fruit pies, the crispness of the crust is a performance characteristic; type and proportions of ingredients, cooking time and temperature are design parameters; variation in the quality of the cooking fat is a random input. In a distillation column (Figure 5.11), the compositions of the tops and bottoms products are performance characteristics; column temperature, column pressure, feed flow rate, reflux flow rate are design parameters; feed composition and steam pressure are random input variables.

5.4.2 Robust design

Figure 5.8 shows the general principles of robust design.[4] These principles can be applied to both product and process design and to both initial 'paper' designs and to prototypes.

The first stage is to specify numerical values of the target vector $\{x_T\}$ and the limit vectors $\{LSL\}$, $\{USL\}$ in the quality specification. The next stage is to build the quality model:

$$\{x\} = f[\{p\}, \{r\}]$$

of the product or process. Initially a computer simulation is the most convenient form of this. The designer must then specify the random variable input vector $\{r\}$, *i.e.* what variables have a significant effect on $\{x\}$ and the likely range of each variable. For example, $\{r\}$ for a pressure transmitter should be based on: ambient temperature −20 to

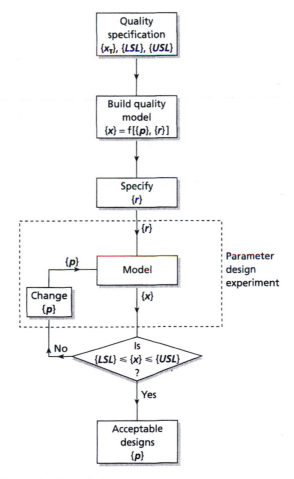

Figure 5.8   Principles of robust design

+30° C, humidity 30 to 70%, supply voltage 10 to 20 V. The next stage is a **parameter design experiment** (Section 5.4.3). Here values of $\{x\}$ are calculated for a range of values of the design parameter vector $\{p\}$. The resulting values of $\{x\}$ are then compared with the specification limits $\{LSL\}$, $\{USL\}$. Acceptable designs are those with values of $\{p\}$ such that $\{x\}$ lies between $\{LSL\}$, $\{USL\}$.

5.4.3 Parameter design experiment

A parameter design experiment is defined by two matrices, the **design parameter matrix** and the **random variable matrix**.

We consider an example where there is a single performance characteristic x, *i.e.*:

$$x = f[\{p\}, \{r\}] \tag{5.7}$$

In our example there are four design parameters p_1, p_2, p_3 and p_4, *i.e.* $\{p\} = p_1, p_2, p_3, p_4$. Each of these parameters can take three values, *e.g.* p_1 can take values p_{11}, p_{12}, p_{13}; p_2 can take values p_{21}, p_{22}, p_{23}, *etc*. This means that there are $4 \times 3 = 12$ different p values and $3^4 = 81$ different possible combinations of these values, *i.e.* 81 different $\{p\}$ vectors. This corresponds to 81 different product designs and 81 different test runs. Taguchi recommends the use of orthogonal arrays to reduce the number of designs tested to a more manageable number, *e.g.* from 81 to 9.[5,6] Figure 5.9 shows the design parameter matrix for this example; this has been constructed using orthogonal arrays. It is a 9×4 matrix; each column corresponds to one of the four design parameters, and each row corresponds to one of the nine product or process designs to be tested.

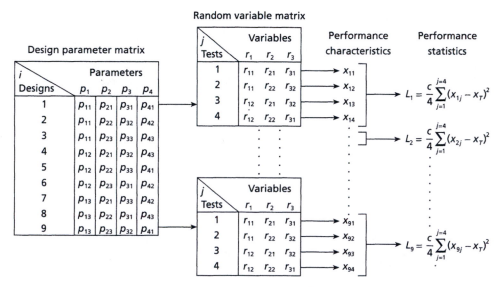

Figure 5.9 Parameter design experiment (after Kacker[4])

The random variable matrix now needs to be defined. Suppose in our example there are three random variables r_1, r_2 and r_3, *i.e.* $\{r\} = r_1, r_2, r_3$. Each of these can take two values, *e.g.* r_1 can take values r_{11}, r_{12}. This means that there are $3 \times 2 = 6$ different r values and $2^3 = 8$ different possible combinations of these values, *i.e.* eight different possible $\{r\}$ vectors. The use of orthogonal arrays reduces the number of $\{r\}$ vectors to be used in the test from 8 to 4. The random variable matrix shown in Figure 5.9 is 3×4; each column corresponds to one of the three random variables, and each row corresponds to one of the four combinations to be tested.

The complete parameter design experiment involves making repeat tests on each row i of the design parameter matrix; each repeat test corresponds to a row j of the random variable matrix. Thus in this example, each of the 9 rows of the design matrix is tested at each of the 4 rows of the random variable matrix making 36 tests in total. The performance characteristic x is measured or calculated in each test giving a total of 36 values x_{ij}, $i = 1, \ldots, 9, j = 1, \ldots, 4$. In general, if the design parameter matrix has m rows and the random variable matrix n rows, the parameter design experiment will consist of $m \times n$ tests and yield $m \times n$ values of the performance characteristic.

The ith design is therefore **acceptable** if all corresponding values of x_{ij}, for that value of i, lie within the specification limits LSL, USL, *i.e.*

$$\text{LSL} \leqslant x_{ij} \leqslant \text{USL} \quad j = 1, 2, 3, 4 \tag{5.8}$$

Thus the 5th design is acceptable if x_{51}, x_{52}, x_{53}, x_{54} all lie within the limits LSL, USL. The process should be repeated for all designs i, $i = 1, \ldots, 9$, to see which designs are acceptable and which are not.

5.4.4 Optimal design

The above designs, while acceptable, are not necessarily **optimal**. We must then decide whether the design should be optimal from the manufacturer's or from the customer's/ user's point of view. Optimal design for the manufacturer should maximize product popularity and the number of units sold. Optimal design for the user should minimize his costs; this is the case we will consider. An optimal design is then one which minimizes the user's costs, *i.e.* the user's loss function $l(x)$, *i.e.* one with x as close as possible to x_T.

For a given design parameter vector $\{p\}$, the effect of changing the random variable vector $\{r\}$ is to cause the performance characteristic x to deviate randomly from the target value x_T. Assuming that the user's loss function $l(x)$ has a minimum value at $x = x_T$, then these random variations in x cause corresponding increases in $l(x)$. The **performance statistic** L_i is the mean value of the user's loss, for the ith row of the parameter design matrix (*i.e.* for the ith design), averaged over the n rows of the random variable matrix, *i.e.*

$$L_i = \frac{1}{n} \sum_{j=1}^{j=n} l(x_{ij}) \tag{5.9}$$

where x_{ij} is the value of the performance characteristic corresponding to the ith row of the design parameter matrix and the jth row of the random variable matrix, $i = 1, \ldots, m$, $j = 1, \ldots, n$. If we assume the quadratic loss function of Equation 5.4, then the ith performance statistic becomes:

$$L_i = \frac{c}{n} \sum_{j=1}^{j=n} (x_{ij} - x_T)^2 \qquad (5.10)$$

Figure 5.9 shows the calculation of the performance statistics L_1 to L_9 corresponding to each of the nine designs to be tested. The optimal design is the one with the lowest value of L.

Example 5.2

Table 5.4 shows values of continuous performance characteristics obtained in a parameter design experiment. Each of nine designs i underwent four tests j to give 36 values x_{ij}.

Table 5.4 Data for Example 5.2

Designs i	Test 1	Test 2	Test 3	Test 4
1	10.2	9.7	10.1	9.9
2	9.8	10.3	10.0	10.1
3	10.1	10.0	9.9	10.0
4	10.0	9.6	9.8	10.3
5	9.9	10.3	10.1	10.1
6	9.7	9.9	10.2	10.0
7	9.8	10.2	9.9	10.1
8	9.9	10.4	10.0	9.7
9	10.2	9.8	9.8	10.2

Product data:
Target value $x_T = 10.0$
Upper spec. limit   USL $= 10.2$
Lower spec. limit   LSL $= 9.8$
Loss function constant $c = 8000$

Using the data given:

(a) Decide which of the designs is acceptable.
(b) Decide which of the acceptable designs is optimal.

Solutions

(a) Want $9.8 \leqslant x_{ij} \leqslant 10.2$ for $j = 1, 2, 3, 4$
This is satisfied by designs $i = 3, 7, 9$

(b) Calculate performance statistic L_i:

$$L_i = c/4 \sum_{j=1}^{4} (x_{ij} - x_{\mathrm{T}})^2$$

for each of the three acceptable designs 3, 7, 9.

$L_3 = 8000/4 \{(10.1 - 10.0)^2 + (10.0 - 10.0)^2 + (9.9 - 10.0)^2$
$+ (10.0 - 10.0)^2\}$
$= 2000 \{(0.1)^2 + (0.1)^2\} = \mathbf{40}$

$L_7 = 8000/4 \{(9.8 - 10.0)^2 + (10.2 - 10.0)^2 + (9.9 - 10.0)^2$
$+ (10.1 - 10.0)^2\}$
$= 2000 \{(0.2)^2 + (0.2)^2 + (0.1)^2 + (0.1)^2\} = \mathbf{200}$

$L_9 = 8000/4 \{(10.2 - 10.0)^2 + (9.8 - 10.0)^2 + (9.8 - 10.0)^2$
$+ (10.2 - 10.0)^2\}$
$= 2000 \{(0.2)^2 + (0.2)^2 + (0.2)^2 + (0.2)^2\} = \mathbf{320}$

The optimal design is therefore 3.

5.4.5 Quality design review

The main requirement is that there are independent checks to ensure that the design, defined by the parameter design vector $\{p\}$, meets the quality specification. The ideal situation is that the review team works closely with the design team as the project progresses, rather than carrying out the review when the design is complete. The reviewer can then make suggestions for improvement at appropriate stages in the design, for example trying a different set of random input variables $\{r\}$ in the robust design. All design changes resulting from this review process should be controlled and clearly documented. Again there should be independent checks that any redesign also meets the quality specification. The culmination of the review process is the design review board, which includes both design and review teams: here the decision is taken to approve the final design.

The project can then progress, if necessary, from 'paper' design to prototype design. A robust design can then be carried out on the prototype. Here a number of prototypes, each with different design parameters $\{p\}$, are built; each one is tested over a specified set of values of the random input vector $\{r\}$. Again there should be independent checks to ensure that the chosen protype design meets the quality specification.

As well as carrying out the **product** design, we may also need to carry out a design of the **process** to manufacture the product. Robust methods can also be used in process design. The random input variable vector $\{r\}$ for the process will be made up of:

(a)   Random variations in input components and materials. Examples are random variations in the characteristics of resistors, capacitors and transistors in an electrical circuit, random variations in the quality of barley, yeast, hops and water in beer production. The design parameter vector $\{p\}$ for the product can be used to identify these random variations, *e.g.* the resistance of key electrical components.

(b)   Random variations in the process itself. Examples are variations in oven temperature, pressure and humidity in a baking process; variations in the alignment of cutting tool and the workpiece in a milling machine.

If we assume a single continuous performance characteristic x for the product, then by testing each process design $\{p\}$ over the specified set of values of $\{r\}$, a set of values of x for each design can be found. These sets of values can be used to:

(a)   Identify acceptable designs. If each value x in the set lies between LSL and USL (Equation 5.8), then the design is acceptable.

(b)   Calculate potential capability index C_P for each design. This is a measure of the capability of a process to meet the product specification and is given by (Section 1.3.2):

$$C_P = \frac{\text{USL} - \text{LSL}}{6\sigma} \tag{5.11}$$

Here σ is the standard deviation of the product performance characteristic and can be estimated, for each design, from the above sets of values of x.

Finally, independent checks of the process design must be carried out.

5.5   Design for reliability

Here the objective is to design a given product or system which meets the target failure rate λ_T (or target reliability R) under the environmental conditions specified. The assumption is made that all components and elements are operating in the useful life region where failure rate is constant with time (Section 2.6.1).

5.5.1 General principles of design for reliability

The following general principles should be observed.

(a)   **Element/component selection** Only elements/components with well-established failure rate data/models (Chapter 4) should be used. Furthermore, some technologies are inherently more reliable than others. For example, solid state switching devices are more reliable than electromechanical reed relays, inductive displacement transducers are more reliable than the resistive potentiometer type.

(b) **De-rating** In Section 4.3.4 **stress** was defined as any variable which when applied to an element or component tends to increase failure rate, examples are mechanical stress and voltage. **Strength** was similarly defined as any property of the element or component which resists the applied stress, examples are elastic limit, rated voltage. In order to reduce failure rate strength should exceed stress by an adequate **safety margin** (Equation 4.6). Thus in a mechanical element a safety margin of better than about 5.0 should be used; in an electronic circuit the voltage stress ratio (Equation 4.7) should be kept well below 0.7.

(c) **Environment** In Sections 4.3.3 and 4.4 we saw how component/element failure rate is critically dependent on the environment and we introduced an **environmental correction factor** π_E into the failure rate model equation. The first step is to define exactly the environment of the element and then estimate the corresponding value of π_E. In the case of a harsh environment with a high value of π_E, elements/components with a high quality level, which are capable of withstanding the environment, should be used. These will have a low quality factor π_Q which will compensate for the high value of π_E and achieve the desired failure rate.

(d) **Minimum complexity** We saw in Section 3.2 that for a series system, the system failure rate is the sum of the individual component/element failure rates. Thus the number of components/elements in the system should be the minimum required for the system to perform its function. In electronic systems reliability can be improved by using integrated circuits which replace many hundreds or thousands of basic devices. The failure rate of the integrated circuit is generally less than the sum of the failure rates of the devices it replaces.

(e) **Redundancy** In Section 3.3 and Example 3.1, we saw that the use of several identical elements/systems connected in parallel increases the reliability of the overall system. Redundancy should be considered in situations where either the complete system or certain elements of the system have too high a failure rate.

(f) **Diversity** The problem of common mode failure was discussed in Section 3.7; here a fault can occur which causes more than one element in a system to fail simultaneously. One example is an electronic system where several of the constituent circuits share a common power supply; failure of the power supply causes all of the circuits to fail. If the probability of common mode failure limits the reliability of the overall system, then the use of equipment diversity should be considered. Here a given function is carried out by two systems in parallel, but each system is made up of different elements with different operating principles. One example is a temperature measurement system made up of two subsystems in parallel; one electronic and one pneumatic.

(g) **Calculation of system reliability** Once the components/elements have been chosen and their configuration in the system/product decided, then the overall system/product reliability can be calculated using Equations 3.1, 3.5, 3.8 and

3.12 for series, parallel and series–parallel systems. If there are multiple components of each type, all connected in series, then the failure rate of the overall system is given by Equation 4.11; this is called **parts count analysis**. Table 4.8 gives an example of this analysis for an electronic square root extractor module. The reliability/failure rate calculated for the overall product should then be compared with the target value; if the target value is not met then the design should be adjusted until the target figure is reached.

5.5.2 Fault tree analysis (FTA)

Until now we have assumed that a component, element or system can only be in one of two states: operating or failed. However, many components/elements and systems can fail in more than one way, *i.e.* have more than one **failure mode**. For example, an electronic component such as a transistor or diode may fail open circuit (maximum resistance) or short circuit (minimum resistance), *i.e.* has at least two failure modes. If the component is part of a more complex circuit, for example the electronic square root element of Table 4.8, then open circuit failure of a given component will cause a certain element failure mode and short circuit failure of that component will cause a different failure mode. Let us assume that open circuit failure of a given transistor causes the element output to rise to a high value and short circuit failure causes the element output to fall to a low value. We further assume that the square root element is itself part of a control loop which controls the flow rate of fluid in a pipe. If the element fails to a high value then the controller closes the valve to reduce the flow; if the element fails to a low value then the controller opens to increase the flow. The two transistor failure modes therefore result in two different control loop failure modes, low flow or high flow. Depending on circumstances, one failure mode will correspond to a fail-safe failure where the plant is moved to a safe condition and one mode will correspond to a fail-danger failure where the plant is moved to a dangerous condition. If the fluid is a heating medium, then high flow failure is fail-danger and low flow failure is fail-safe, so that open circuit transistor failure is fail-safe and short-circuit failure is fail-danger. If the fluid is a cooling medium, then high flow failure is fail-safe and low flow failure is fail-danger; here open circuit transistor failure is fail-danger and short-circuit failure is fail-safe.

The system should therefore be designed so that the probability of fail-danger is minimized. Two main techniques are available to the designer; these are fault tree analysis (FTA) and failure mode and effect analysis (FMEA).

Fault tree analysis is a systematic way of identifying all possible faults that could lead to system fail-danger failure. It is a 'top-down' or 'backward logic' method; starting from the fail-danger failure as the top event, a logic diagram is constructed showing all possible combinations of faults and conditions that could cause the top event. This logic diagram is built up from a number of AND and OR gates; Figure 5.10 shows the use of AND and OR gate symbols to represent two different ways of combining probabilities and is based on Equations 1.26 and 1.28.

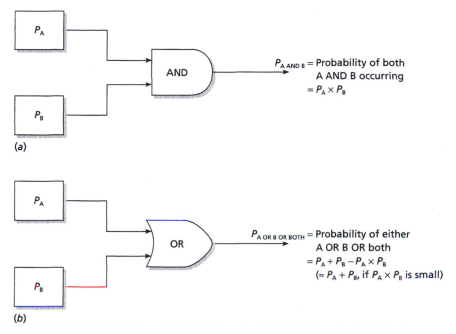

(a)

$P_{A \text{ AND } B}$ = Probability of both A AND B occurring = $P_A \times P_B$

(b)

$P_{A \text{ OR } B \text{ OR BOTH}}$ = Probability of either A OR B OR both = $P_A + P_B - P_A \times P_B$ ($\approx P_A + P_B$, if $P_A \times P_B$ is small)

Figure 5.10   Methods of combining probabilities

Example 5.3

Figure 5.11 shows a simplified process and instrument line diagram for a distillation column. A hazardous situation is created if the flow rate of steam to the reboiler goes high; this causes a high flow rate of vapour up the column producing a high pressure which could cause the vessel to rupture. The temperature control loop consists of a platinum resistance thermometer (PRT), a transmitter (which converts resistance change to a 4–20 mA current signal), a controller, a current-to-pneumatic converter and a control valve.

The control valve is shut at minimum air pressure (3 psig) and fully open at maximum air pressure (15 psig). The plant is protected by a pressure trip system consisting of a pressure switch and a three-way solenoid valve located in the air line between the converter and the control valve.

Under normal conditions the pressure switch is closed, the solenoid valve is energized, the vent port shut and the air supply to the control valve is normal. If the column pressure exceeds the trip setting, the pressure switch opens to de-energize the solenoid and the vent port opens, this causes the pressure in the control value bonnet to fall to zero and the valve to close, thus shutting off the flow of steam and making the plant safe. The trip system therefore increases the probability that any failure will be a fail-safe failure. There is still a non-zero probability of a fail-danger failure caused by the control valve going fully open and giving a high flow. This is calculated by identifying all possible cause of the event, assigning a probability to each and then combining these probabilities. The fault tree diagram and calculation of the overall fail-danger probability F is shown in Figure 5.12.

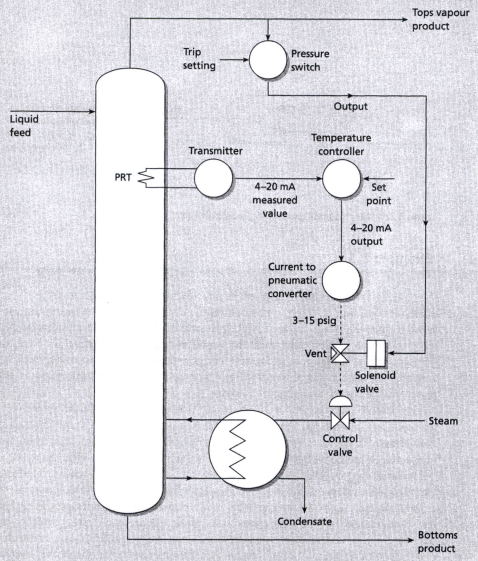

Figure 5.11 Process and instrument line diagram for distillation column

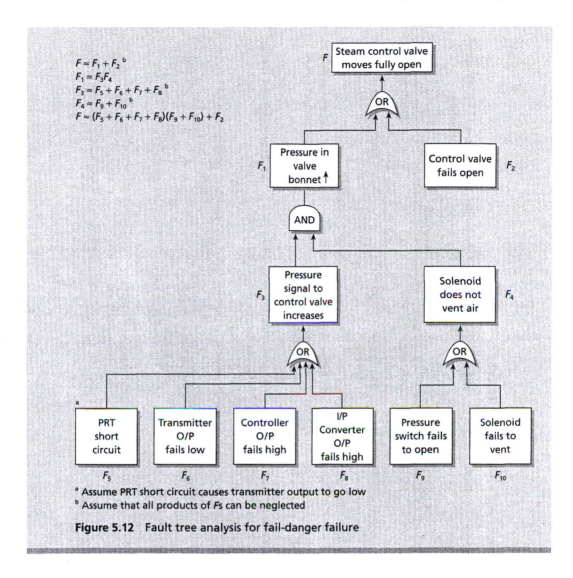

$F \approx F_1 + F_2$ [b]
$F_1 = F_3 F_4$
$F_3 \approx F_5 + F_6 + F_7 + F_8$ [b]
$F_4 \approx F_9 + F_{10}$ [b]
$F \approx (F_5 + F_6 + F_7 + F_8)(F_9 + F_{10}) + F_2$

[a] Assume PRT short circuit causes transmitter output to go low
[b] Assume that all products of Fs can be neglected

Figure 5.12   Fault tree analysis for fail-danger failure

5.5.3   Failure mode and effect analysis (FMEA)

In fault tree analysis **all** possible failure modes that could lead to the top event are identified. If there are a large number of components in a given element, then FTA may not be practicable. In this case **failure mode and effect analysis** may be more useful. Here all **significant** component failure modes are listed and their effect on the element, *i.e.* whether fail-safe or fail-danger, evaluated and a failure rate assigned to each failure mode. The element fail-danger/fail-safe failure rate λ_D, λ_S is then the sum of the component fail-danger/fail-safe failure rates (Equation 3.5). The corresponding element fail-danger and fail-safe probabilities are then given by (Equation 2.16):

Table 5.5 Examples of failure mode and effect analysis (FMEA)

Element	Component	Fault	Effect	Failure rate (faults/y)		Method of elimination or reduction
				Dangerous	Safe	
Platinum resistance thermometer	Resistance element	Short circuit	Transmitter O/P low – fail-danger	0.05		Two elements in parallel
	Resistance element	Open circuit	Transmitter O/P high – fail-safe		0.35	Two elements in parallel
	Total			0.05	0.35	
Current-to-pneumatic converter	Input circuit	Short circuit	O/P goes high – fail-danger	0.05		Fit fuse to ensure open circuit failure
	Input circuit	Open circuit	O/P goes low – fail-safe		0.25	
	Output pneumatics	Blocked nozzle	O/P goes high – fail-danger	0.02		Ensure clean air supply
	Output pneumatics	Ruptured bellows	O/P goes low – fail-safe		0.15	
	Total			0.07	0.40	
Pressure switch	Spring	Fracture	Switch fails to open – fail-danger	0.01		
	Bellows	Rupture	Switch opens – fail-safe		0.20	
	Pivot screws	Loosen	Switch fails to open – fail-danger	0.04		
	Microswitch	Random	20% danger 80% fail	0.01	0.04	
	Total			0.06	0.24	

$$F_D = 1 - e^{-\lambda_D t}, \quad F_S = 1 - e^{-\lambda_S t} \tag{5.12}$$

Table 5.5 shows examples of FMEA for three elements of Figure 5.11, platinum resistance thermometer, current-to-pneumatic converter and pressure switch. In situations where a given element has to be designed to meet target values for λ_D and λ_S, then FMEA highlights component failure modes which have the greatest effect and where corrective action should be taken.

Example 5.4

Figure 5.13 shows a schematic diagram of the control system for a domestic gas boiler. A system fail-danger failure corresponds to a gas explosion due to unburnt gas being ignited. A system fail-safe failure corresponds to the boiler being automatically shut down unnecessarily. Table 5.6 shows the FMEA for the system; the effect of each failure mode is analysed and overall fail-safe and fail-danger failure rates are calculated.

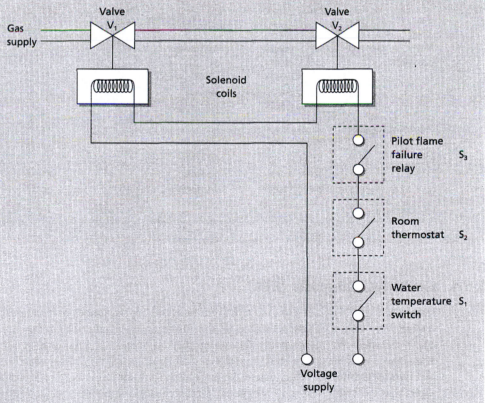

Figure 5.13 Domestic gas boiler control system

Table 5.6 FMEA analysis for domestic gas boiler

Element	Failure Mode	Effect	Fail-danger failure rate yr^{-1}	Fail-safe failure rate yr^{-1}
Gas valve 1	stuck closed	fail safe	–	2×10^{-5}
	stuck open	fail danger	2×10^{-5}	–
	through leakage	fail danger	3×10^{-4}	
Gas valve 2	stuck closed	fail safe	–	2×10^{-5}
	stuck open	fail danger	2×10^{-5}	–
	through leakage	fail danger	3×10^{-4}	
Solenoid coils	short circuit	fail safe	–	10^{-5}
	open circuit	fail safe	–	10^{-5}
	earth fault	fail safe	–	10^{-5}
Pilot flame relay	stuck closed	fail danger	2×10^{-3}	–
	stuck open	fail safe	–	2×10^{-3}
	short circuit	fail danger	2×10^{-3}	–
	open circuit	fail safe	–	2×10^{-3}
	earth fault	fail danger	1×10^{-3}	–
Water temperature switch	stuck closed	fail danger	1×10^{-3}	–
	stuck open	fail safe	–	1×10^{-3}
	short circuit	fail danger	2×10^{-3}	–
	open circuit	fail safe	–	2×10^{-3}
	earth fault	fail danger	1×10^{-4}	–
Room thermostat	stuck closed	fail safe	–	1×10^{-3}
	stuck open	fail safe	–	1×10^{-3}
	short circuit	fail safe	–	2×10^{-3}
	open circuit	fail safe	–	2×10^{-3}
	earth fault	fail danger	1×10^{-2}	–
Total failure rates			1.87×10^{-2}	1.31×10^{-2}

5.5.4 Event tree analysis (ETA)

Event trees use 'forward logic' to define all possible consequences of a failure within a system. The failure is first identified as the initiating event and the probability of this event specified. We then identify the events that occur next in time in response to this event; there may be several and we assign a probability to each. In many cases there are two **complementary** events, where the sum of the probabilities is 1: an example is an alarm that either does or does not go off. For each of these events, we then identify the next events in time and again assign a probability to each. The process is continued until the tree is built up. The process concludes by identifying a number of final outcomes, for example whether an accident does or does not occur. The probability of each outcome can then be found by:

(a) first identifying the sequence of events which leads to the outcome and
(b) multiplying the probabilities of all the events in the sequence together (corresponding to an AND combination of probabilities).

Example 5.5

Figure 5.14 shows an event tree associated with the escape of an odourless, poisonous gas in a chemical plant. The probability of this initiating event and all subsequent events are shown in the figure. The probability of each of the outcomes E_1, E_2, E_3, E_4, E_5 are calculated and the total probabilities of either an accident or no accident found.

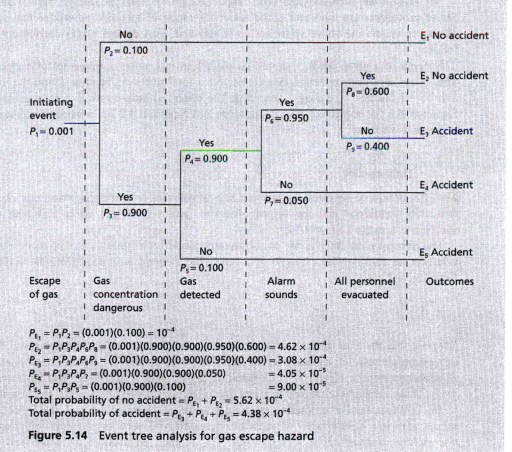

$P_{E_1} = P_1 P_2 = (0.001)(0.100) = 10^{-4}$
$P_{E_2} = P_1 P_3 P_4 P_6 P_8 = (0.001)(0.900)(0.900)(0.950)(0.600) = 4.62 \times 10^{-4}$
$P_{E_3} = P_1 P_3 P_4 P_6 P_9 = (0.001)(0.900)(0.900)(0.950)(0.400) = 3.08 \times 10^{-4}$
$P_{E_4} = P_1 P_3 P_4 P_7 = (0.001)(0.900)(0.900)(0.050) \quad\quad = 4.05 \times 10^{-5}$
$P_{E_5} = P_1 P_3 P_5 = (0.001)(0.900)(0.100) \quad\quad\quad\quad = 9.00 \times 10^{-5}$
Total probability of no accident $= P_{E_1} + P_{E_2} = 5.62 \times 10^{-4}$
Total probability of accident $= P_{E_3} + P_{E_4} + P_{E_5} = 4.38 \times 10^{-4}$

Figure 5.14   Event tree analysis for gas escape hazard

5.5.5 Reliability design review and testing

The objective here is to design a product or system, which meets the target failure rate λ_T under the range of environmental conditions specified. We saw in Section 4.3 that failure rate, while nominally constant with time during the useful life region, will depend on temperature, environmental variables and stress. All of these quantities can vary randomly, causing failure rate to vary randomly about a mean value. A more realistic reliability specification would state that all products with λ between a lower specification limit LSL and an upper specification limit USL are acceptable for sale.

The main requirement of the **reliability design review** is that there are independent checks to ensure that the 'paper' design meets the above apecification. This is best done by the review team working with the design team; for example, ensuring that the effects of all likely variations in environmental conditions and component failure rates have been considered. This process concludes with the design review board which makes the final decision to approve the design.

The project can then progress from 'paper' design to prototype design. Here a number of prototypes are built and tested under the range of specified environmental conditions to ensure the above specification is satisfied. This **design** test is carried out as follows.

N items of a given product should be placed on test, under a given set of temperature, stress and environmental conditions, for a total observation time T and the total number of failures N_F during T recorded. The observed failure rate $\bar{\lambda}$ is then given by Equation 4.1 for non-repairable components and Equation 4.2 for repairable elements; in general we have:

$$\bar{\lambda} = \frac{N_F}{\text{Total \textbf{up} time}} \tag{5.13}$$

The test should be repeated at different sets of relevant temperature, stress and environmental conditions and the set of observed failure rates $\{\bar{\lambda}\}$ compared with the specification range LSL to USL.

The above failure rates $\bar{\lambda}$ are only estimates based on a limited amount of information. Let us consider the example of a repairable element with negligible down time; Equation 4.2 gives:

$$\bar{\lambda} = \frac{N_F}{NT} \tag{5.14}$$

Suppose that the true value of failure rate λ for the element is 0.5 faults/year and that three separate tests are carried out; for each test $N = 20$ and $T = 1.0$ year. The values of N_F obtained were 8, 10 and 12 corresponding to estimates $\bar{\lambda}$ of 0.4, 0.5 and 0.6 respectively. These variations are sampling errors due to the limited size of N_F; if NT was increased to say 200 then the corresponding estimate $\bar{\lambda}$ would be closer to the true value. For elements with low failure rate, NT must be increased to a very large value before even one failure is observed; if $\lambda = 0.1$ and $N = 20$ then on average we must wait for $T = 0.4$ years before there is a 95% probability of up to two failures occurring. One solution to this problem is to increase λ artificially by increasing the stress level on the

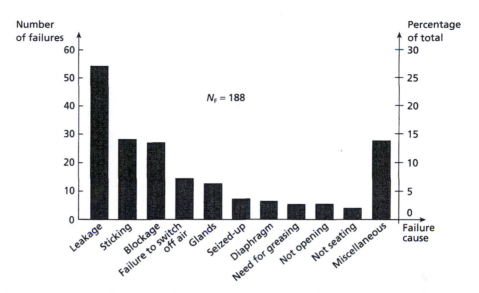

Figure 5.15 Pareto distribution for causes of control valve failure

element or component. This approach is satisfactory provided that the highest stress level is still within the valid range of the failure rate–stress relationship. The procedure of assigning **confidence levels** to a given estimate $\bar{\lambda}$ is discussed in Section 5.7.4.

Once it has been proved that the prototype design meets the specification, the next stage is usually to build production models. Testing of production models against the reliability specification is referred to as **qualification testing**. This usually involves testing over the full range of environmental conditions and field trials in the normal working environment.

As well as providing overall failure rate data $\bar{\lambda}$ for the product, the above tests should also provide data on:

(a)   the basic individual causes of failure and the components responsible
(b)   the number of failures due to each basic cause as a percentage of the total number of overall failures.

This data is displayed using a **Pareto distribution**, which is a plot of how a total number of failures N_F is distributed among all possible individual causes. Figure 5.15 shows a Pareto distribution for a total of 188 overall failures in a given type of control valve.[7] The graph shows both the number of failures due to each cause and the corresponding percentages; the highest are on the left, lowest on the right. We note that 58% of the failures are due to three basic causes: leakage, sticking and blockage.

5.6   Quality in manufacture

Figure 5.16 is a representation of a manufacturing process. Random variations in the performance characteristic x for the output product occur due to random variations in both:

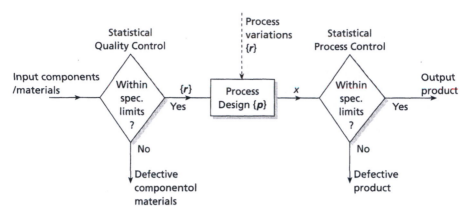

Figure 5.16   Manufacturing Process with Quality Control

(a)   input materials and components
(b)   the manufacturing process itself.

These random variations make up the random variable input vector $\{r\}$.

Assuming that a robust design has been carried out on the process, then the chosen design (specified by design vector $\{p\}$) will give x within the specification limits LSL, USL, for all values of $\{r\}$ considered at the design stage. If, during manufacture, these random variations are greater than expected, then there will be a greater range of values of $\{r\}$ than allowed for in the design. This means that it is now likely that x will fall outside the specification limits. To avoid this probem there must be quality control on both input components/materials and the output product.

The first step is to establish a clearly defined specification for input components and materials. The design values of $\{r\}$ should be used to set corresponding lower and upper specification limits for component/material performance characteristics. The performance characteristics should then be tested against these limits using **statistical quality control** techniques (Section 1.3) and defective items rejected. The test procedures to be used should be agreed between component/materials suppliers and the customer. **Sampling** techniques (Section 5.7.4) are essential if there are a large number of components; again these must be agreed between supplier and customer.

All manfacturing processes are subject to random variations. The most common sources of variability are manufacturing operations involving humans, especially repetitive, boring or unpleasant tasks. The best method of eliminating human variability is to automate the tasks; examples are computer controlled machine tools, robot spot welding and painting, automatic component placement and soldering in electronic circuits. Automated processes also show variability, for example small fluctuations in temperature in a thermostatically controlled oven, but here the variations are usually smaller. **Statistical process control** techniques (Section 1.3) should be used where possible. Here online measurements of product performance characteristic x are made and compared with corresponding specification limits LSL, USL: defective products are rejected. **Quality control charts** (Section 1.3.3) are especially useful here.

5.7 | Testing of the finished product

The finished product must be inspected and tested to ensure that it conforms to the specification; this is often referred to as **demonstration** or **acceptance** testing.

5.7.1 Quality test

A quality test is required to demonstrate that the finished product:

(a) Meets the specification for **binary** performance characteristics or **attributes**. Thus a finished car should be checked to ensure that all equipment and fittings are both present and working correctly.

(b) Meets the specification for **continuous** performance characteristics. Here the set of performance characteristics $\{x\}$ should be measured, for all values of random input variables $\{r\}$ specified at the design stage, to check that they are within the specification limits $\{LSL\}$, $\{USL\}$. Thus the carbon monoxide level in the car exhaust should be measured to ensure it is less than 1%.

5.7.2 Reliability test

A reliability test is also required to demonstrate that the finished product meets the reliability specification defined by target failure rate λ_T (with specification limits LSL, USL) under all specified environmental conditions. A full reliability test may consist of three different types of activities: an example is the US Military Specification MIL-STD-883 for microelectronic devices.[8]

(a) **Measurement of failure rate** The failure rate of individual items or random samples of the product should be measured under the range of temperature, stress and environmental conditions defined by the specification. The test could therefore involve testing at different levels of temperature, humidity, mechanical vibration, mechanical shock, operating voltage and salinity.

(b) **'Burn-in'** Here all items of the product are subject to levels of temperature, stress and environmental conditions which are more severe than the specification. This should accelerate the failure of weak or defective components so that they fail during the burn-in test and are therefore weeded out before despatch to the customer. The aim is to eliminate the **early failure region** of the 'bathtub' curve (Section 2.6) so that the product despatched to the customer is in the **useful life region** which is characterized by low constant failure rate. Burn-in is particularly useful with electronic components and assemblies, where accelerated failure is mainly produced by high levels of temperature, humidity and voltage.

(c) **Screening** This is a range of other tests which can be used to identify defects; examples are visual inspection, constant acceleration and measurement of electrical parameters. Again this is mainly applicable to electronic components and assemblies.

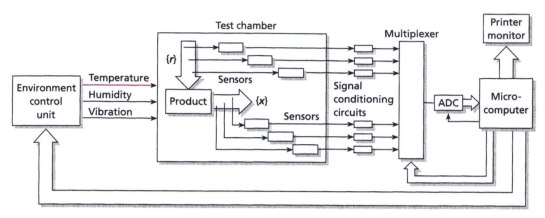

Figure 5.17   Automatic test equipment (ATE)

5.7.3 Automatic test equipment (ATE)

Figure 5.17 shows the general form of a system for automatic testing. Items of the product are located inside a test chamber where the environment can be altered in a controlled way. Relevant product performance characteristics $\{x\}$ and environmental variables $\{r\}$ are measured with appropriate sensors and signal conditioning circuits convert the sensor outputs to a common signal range, *e.g.* 0 to 5 V (Section 6.5.2). These analogue signals are input to a multiplexer and the multiplexed signal passes to a single analogue-to-digital converter and then to a parallel input port of a microcomputer. The computer calculates the measured values of $\{x\}$ and $\{r\}$ and outputs the data to a monitor and printer via a serial/parallel output port. The computer also controls the operation of the multiplexer, analogue-to-digital converter and the environmental control unit which adjusts the environment inside the chamber, *e.g.* temperature, humidity and vibration. The computer can thus be programmed to execute and record the entire test programme automatically.

5.7.4 Sampling schemes

Ideally all items of the product should pass the above quality and reliability tests before despatch to the customer; similarly all bought-in components should pass these tests before being used in manufacture. Items are often delivered in lots or batches of a given number and it is often impossible or uneconomic to test every single item in the lot. A more practical method is to take random samples from the lot and test each item in the sample. With sampling there is always a non-zero probability that an unsatisfactory lot will be passed; this is called the **customer's risk**. Similarly there is also a non-zero probability that a good lot will be rejected; this is called the **producer's risk**. Sampling schemes are used to find the sample size and the number of allowable sample defects corresponding to a given customer's/producer's risk.

Single sampling schemes

A single sampling scheme is defined by (Figure 5.19(a)):[9]

Test a sample of n items from a lot of size N
If $0 \leqslant$ number of defects $\leqslant c -$ accept the lot
If number of defects $> c -$ reject the lot

The probability that the lot is accepted, *i.e.* the probability of between 0 and c defects, is given by the **cumulative Poisson** probability distribution (Appendix A):

$$P_{0,c} = \exp(-z) \sum_{k=0}^{k=c} \frac{z^k}{k!} \quad \text{cumulative Poisson probability distribution} \tag{5.15}$$

where $P_{0,c}$ = probability of getting between 0 and c defects in sample of size n
   $z = n \times p$
   p = fraction of defectives in entire lot

Thus when:

$$c = 0, P_{0,0} = e^{-z} \tag{5.16}$$
$$c = 1, P_{0,1} = e^{-z}(1 + z) \tag{5.17}$$
$$c = 2, P_{0,2} = e^{-z}(1 + z + \frac{z^2}{2}) \tag{5.18}$$
$$c = 3, P_{0,3} = e^{-z}(1 + z + \frac{z^2}{2} + \frac{z^3}{6}) \tag{5.19}$$

Table 5.7 gives values of $P_{0,0}, P_{0,1}, P_{0,2}, P_{0,3}$ for values of z between 0 and 5. Figure 5.18 shows each probability plotted as a function of z; these are referred to as **operating characteristic curves** (OCC). These curves are used to generate two types of sampling plans:

Table 5.7   Values of cumulative Poisson distribution $P_{0,c}$ for $c = 0, 1, 2, 3$

z	$P_{0,0}$ ($c=0$)	$P_{0,1}$ ($c=1$)	$P_{0,2}$ ($c=2$)	$P_{0,3}$ ($c=3$)
0.00	1.000	1.000	1.000	1.000
0.10	0.905	0.995	1.000	1.000
0.20	0.819	0.982	0.999	1.000
0.30	0.741	0.963	0.996	1.000
0.40	0.670	0.938	0.992	0.999
0.50	0.607	0.910	0.986	0.998
0.60	0.549	0.878	0.977	0.997
0.70	0.497	0.844	0.966	0.994
0.80	0.449	0.809	0.953	0.991
0.90	0.407	0.772	0.937	0.987
1.00	0.368	0.736	0.920	0.981
2.00	0.135	0.406	0.677	0.857
3.00	0.050	0.199	0.423	0.647
4.00	0.018	0.092	0.238	0.433
5.00	0.007	0.040	0.125	0.265

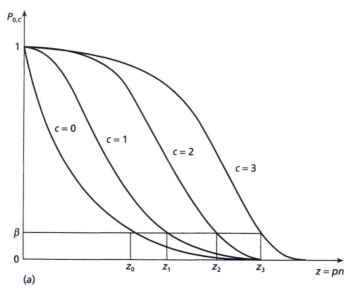

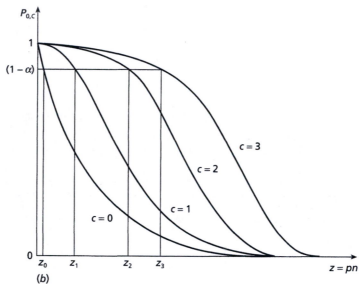

Figure 5.18 Operating characteristic curves and sampling plans: (a) LTPD sampling; (b) AQL sampling

Lot tolerance percent defective (LTPD)
Acceptable quality level (AQL)

LTPD sampling is based on the customer's risk. If the customer's risk is β, then the probability that a lot with p equal to the LTPD specified will be rejected is $1 - \beta$, and the probability that it is accepted is β. Figure 5.18(a) shows that for a given value of β, the values of z_0, z_1, z_2, z_3 corresponding to $c = 0, 1, 2, 3$ defects can then be found. Assuming that LTPD for the lot is known and since $z = n \times$ LTPD, then the values of sample size n corresponding to $c = 0, 1, 2, 3$ are $n_0 = z_0/$LTPD, $n_1 = z_1/$LTPD, $n_2 = z_2/$LTPD, $n_3 = z_3/$LTDP, respectively.

AQL sampling is based on the producer's risk. If the producer's risk is α, then the probability that a lot with p equal to the AQL specified will be rejected is α and the probability that it is accepted is $(1 - \alpha)$. Figure 5.18(b) shows that for a given value of $(1 - \alpha)$, the values of z_0, z_1, z_2, z_3, corresponding to $c = 0, 1, 2, 3$ defects, can then be found. Assuming that AQL for the lot is known and since $z = n \times$ AQL, then the values of sample size n corresponding to $c = 0, 1, 2, 3$ are $n_0 = z_0/$AQL, $n_1 = z_1/$AQL, etc.

Example 5.6

(a) A lot of electronic components has LTPD = 5% and samples of size 100 are taken from the lot. What is the probability that the entire lot will be accepted if acceptance is based on:
   (i) No defects in the sample?
   (ii) 0 or 1 defects in the sample?
   (iii) 0, 1 or 2 defects in the sample?
(b) A lot has AQL = 1% and samples of size 50 are taken from the lot. What is the probability that the entire lot will be rejected if acceptance is based on:
   (i) No defects in the sample?
   (ii) 0, 1, 2 or 3 defects in the sample?

Solutions

(a) LTPD = 5% = 0.05, $n = 100$, $z = $ LTPD $\times n = 5.00$
   (i) From Table 5.7, $P_{0,0} = \beta = 0.007$ when $c = 0$, *i.e.* customer's risk is 0.7% that the lot will be accepted.
   (ii) $P_{0,1} = \beta = 0.040$ when $c = 1$, *i.e.* customer's risk is 4.0%
   (iii) $P_{0,2} = \beta = 0.125$ when $c = 2$, *i.e.* customer's risk is 12.5%
(b) AQL = 1% = 0.01, $n = 50$, $z = $ AQL $\times n = 0.5$
   (i) $P_{0,0} = 1 - \alpha = 0.607$ when $c = 0$, *i.e.* $\alpha = 0.393$, *i.e.* producer's risk is 39.3% that the lot will be rejected.
   (ii) $P_{0,3} = 1 - \alpha = 0.998$ when $c = 3$, *i.e.* $\alpha = 0.002$, *i.e.* producer's risk is 0.2% that the lot will be rejected.

We can also use use Equation 5.15 and Table 5.7 to estimate the observation time T that must elapse before a given degree of confidence can be placed in the observed failure rate λ obtained during a reliability test (Section 5.5.5). If a total of N repairable items are placed on test and N is the number of failures during T, then we set:

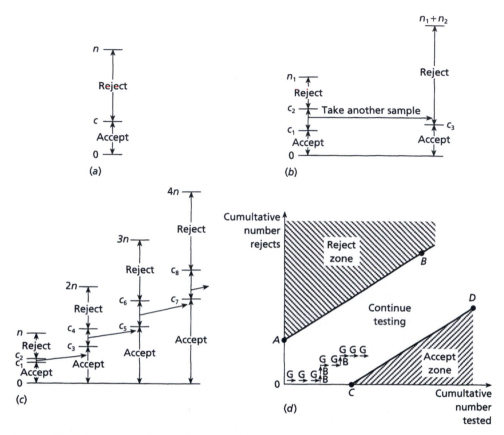

Figure 5.19 Sampling schemes (after Haslehurst[8]): (a) single sampling; (b) double sampling; (c) multiple sampling; (d) sequential sampling

$$z = \lambda NT \quad \text{and} \quad c = N_F \tag{5.20}$$

Suppose $\lambda = 0.1$ per year, $N = 20$ and we need to estimate the time T required for the test so that the failure rate can be demonstrated with 95% confidence for two or fewer failures. From Table 5.7, $z \approx 0.8$, when $P_{0,2} = 0.95$ and $c = N_F = 2$; this gives

$$T = \frac{0.8}{20 \times 0.1} = 0.4 \text{ year}$$

Double sampling scheme

Single sampling can result in a 'sudden death' outcome where the whole lot may be rejected on the evidence of a single sample. In a double sampling scheme, a second sample is taken in order to reinforce the evidence of the first sample; this can be defined by (Figure 5.19(b)):

Test a sample of n_1 items from a lot of size N
  If number of defects $< c_1$ – accept the lot
  If number of defects $> c_2$ – reject the lot
  If $c_1 \leqslant$ number of defects $\leqslant c_2$ – take another sample
Then treat the operation as single sampling with sample size $n_1 + n_2$ and acceptance value c_3.

The total number of items tested will be less than in a statistically equivalent single scheme, but this advantage may be outweighed by the extra administrative costs of running the scheme.

Multiple and sequential sampling schemes

The principle of double sampling can be extended to **multiple sampling**; here a large number of small samples, all of the same size n, are taken until a decision either to accept or reject the entire lot is reached. Figure 5.19(c) shows a possible multiple sampling scheme; the advantage of multiple sampling is that the total number of items tested will be less than in statistically equivalent double schemes, the disadvantage is that it is more complex to administer.

The principle of multiple sampling can be extended until samples containing only a single item, *i.e.* $n = 1$ are tested; this type of sampling is called **sequential sampling**. Figure 5.19(d) shows sequential sampling in graphical form. Here the $0-y$ axis represents the cumulative number of rejects, and the $0-x$ axis the cumulative number of items tested; two parallel lines, drawn through the points A, B, C and D, define three zones Accept, Reject and Continue Testing. Good (G) and Bad (B) items are plotted in the Continue Testing zone as shown using horizontal or vertical lines respectively, until either of the zone boundaries is crossed; then the entire lot is either accepted or rejected. The co-ordinates of A, B, C and D can be found using the Operating Characteristic curves.

Summary

This chapter commenced by discussing Total Quality Management, *i.e.* the methods and culture necessary to achieve quality. It then went on to discuss how quality and reliability affect the profitability of a product. Methods of ensuring quality and reliability at the design and manufacturing stages were then explained. The chapter concluded by discussing testing methods necessary for quality and reliability assurance of the finished product.

References

1. BS 5750 (1987) *Quality Systems, Part One (ISO9001/EN29001): Specification for design/development, production, installation and servicing*, British Standards Institution, London.
2. Smith, D. J. (1988) *Reliability and Maintainability in Perspective*, 3rd edn, Macmillan, Basingstoke, pp. 9–11.

3. Kacker, R. N. (1989) Taguchi's quality philosophy: Analysis and commentary. In K. Dehnad (ed.) *Quality Control, Robust Design and the Taguchi Method*, Wadsworth and Brooks/Cole, California, pp. 3–19.
4. Kacker, R. N. (1989) Off-line quality control, parameter design and the Taguchi method. In K. Dehnad (ed.) *Quality Control, Robust Design and the Taguchi Method*, Wadsworth and Brooks/Cole, California, pp. 51–76.
5. Taguchi, G. (1976–1977) *Experimental Designs*, 3rd edn, 2 vols, Maruzen, Tokyo.
6. Kolarik, W. J. (1995) *Creating Quality: Concepts, Systems, Strategies and Tools*, New York, McGraw-Hill, pp. 435–482.
7. Lees, F. (1976) The reliability of instrumentation, *Chemistry and Industry*, March, p. 198.
8. National Semiconductor Corporation (1987) *The Reliability Handbook*, 3rd edn, NSC, California, pp. 18–33.
9. Haslehurst, M. (1969) *Manufacturing Technology*, English Universities Press, London, pp. 355–62.

Self-assessment questions

5.1 A manufacturer markets three different types of washing machines at quality levels 1, 2, 3. Table Q5.1 gives the following data for each quality level:

Table Q5.1 Data for Question 5.1

Quality	c_Q £	λ_T/yr	c_F £	c_0 £1000s	N 1000s	s £
1	50	1.0	150	1000	20	300
2	100	0.5	250	1000	12	500
3	200	0.2	350	1000	5	700

(a) Variable manufacturing cost per unit c_Q £
(b) Target failure rate λ_T per year
(c) Manufacturer's cost per failure c_F £
(d) Fixed manufacturing cost c_0 £
(e) Number of units manufactured per year N
(f) Selling price per unit s £

Which is the most profitable type of machine for the manufacturer?

5.2 A parameter design experiment was carried out on a semiconductor manufacturing process. Each of five possible designs was tested five times to give the following data for semiconductor energy gap E in electron-volts:

Table Q5.2 Data for Question 5.2

Designs	Test 1	Test 2	Test 3	Test 4	Test 5
1	5.07	5.12	5.10	5.03	5.05
2	4.91	4.87	4.95	4.89	4.96
3	4.82	5.17	5.02	4.78	5.23
4	4.95	5.04	4.98	5.03	4.96
5	4.82	4.93	5.02	5.10	5.19

The loss function associated with deviations in E from the target value of 5.00 eV is:

$$l(E) = 10\ 000\ (E - 5.00)^2$$

Decide which of the designs is optimal.

5.3 This problem refers to the domestic gas boiler control system shown in Figure 5.13. The two gas valves V_1 and V_2 are identical and in hydraulic series; each valve has two possible fail-danger fault conditions:

valve stuck open (probability $F_1 = 10^{-4}$)
leakage flow through valve (probability $F_2 = 10^{-3}$).

Either of these faults can give a fault flow through the valve.

The three switches S_1, S_2, S_3 are identical and in electrical series; each switch has three possible fail-danger fault conditions:

stuck closed (probability $F_3 = 2 \times 10^{-3}$)
short circuit (probability $F_4 = 2 \times 10^{-3}$)
earth fault (probability $F_5 = 10^{-3}$)

Any one of these faults can give a fault current flow through the switch. If a current flows through the circuit, the solenoids are energized and both valves V_1 and V_2 are open.

The probability of the pilot flame failing $F_6 = 10^{-1}$
The probability of a source of ignition $F_7 = 1$

(a) Draw a fault tree associated with the hazard of a gas explosion.
(b) Use the data given to calculate the probability of a gas explosion.

5.4 A rubber product is made by injecting liquid rubber into a mould; the mould contains a metal fixture stamped into a particular shape covered with adhesive and laced into the mould. The product was tested for defects and the results shown in Table Q5.4 obtained:

Table Q5.4 Data for Question 5.4

Type of defect	Number of defects
Bad rubber	91
Poor adhesion	128
Cracks	9
Voids	36
Impurities	15
Cuts	23
Other	12

Use the results to sketch Pareto distributions showing:

(a) the number of defects of each type
(b) the probability of each type of defect

(c)   the cumulative total of defects

(d)   the cumulative probability of defects plotted against the type of defects.

5.5   A lot of mechanical components has LTPD = 10%. The customer's risk, *i.e.* the probability that the lot is accepted, is to be 9%.

(a)   Calculate the minimum sample size if acceptance is based on no defects in the sample (use Equation 5.16).

(b)   Estimate the minimum sample size if acceptance is based on 0 or 1 defects in the sample (Table 5.7).

(c)   If acceptance is based on 0, 1 or 2 defects in the sample, use Equation 5.18 to prove that the minimum sample size should be approximately 55.

5.6   A lot of electrical components has AQL = 5%. The producer's risk, *i.e.* the probability that the lot will be rejected, is to be 4%.

(a)   Calculate the minimum sample size if acceptance is based on no defects in the sample (use Equation 5.16).

(b)   Estimate the minimum sample size if acceptance is based on 0 or 1 defects in the sample (Table 5.7).

(c)   If acceptance is based on 0, 1, or 2 defects in the sample, use Equation 5.18 to prove that the minimum sample size should be approximately 15.

Reliability and availability in maintenance

6.1 | Introduction

This chapter begins by looking at the effects of reliability and maintainability on the total costs incurred by the user during the lifetime of the purchased system. The economics of three maintenance strategies: breakdown, preventive and on-condition are then compared. Each of the three strategies is then discussed in detail.

6.2 | Economics of maintenance

6.2.1 Reliability and the total cost of ownership

We saw in Section 5.3.4 that for the worked example of a transmitter with six different levels of quality, the manufacturer's profit was maximum at quality level 4. Failure rate will generally depend on quality level; the higher the quality level the lower the failure rate and vice versa. Thus for this example, the manufacturer's profit depends on failure rate and has a maximum value at a failure rate λ_M (Figure 6.1); this is the optimum failure rate for the manufacturer. We also saw that the manufacturer's selling price for each unit increases as quality level increases. When a user purchases an item of equipment from a manufacturer and puts it into operation, he incurs an initial cost C_I. This initial cost is made up of the costs of purchase, delivery, installation and commissioning. The cost of purchasing is equal to the manufacturer's selling price and is a major contributor to C_I. User initial cost will therefore decrease as quality decreases, *i.e.* as failure rate increases (Figure 6.1). The user will also have to pay the cost of system failures, *e.g.* lost production and efficiency and the cost of repairing these failures, *i.e.* materials

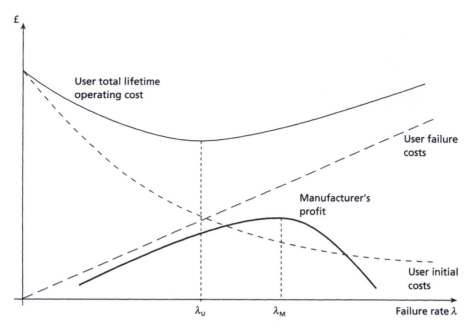

Figure 6.1 User costs as a function of failure rate

and labour costs, throughout the lifetime of the system. This failure cost will increase with failure rate (Figure 6.1). The relevant cost to the user is therefore **total lifetime operating cost** TLOC which is the total cost incurred by the user during the lifetime of the system and is the sum of the above costs, *i.e.*:

$$\text{TLOC} = \text{Initial cost } C_I + \text{Cost of failures and maintenance throughout lifetime} \qquad (6.1)$$

Figure 6.1 shows how TLOC varies with failure rate; we see that TLOC has a minimum at failure rate λ_U. This is the optimum failure rate for the user. We note that λ_U is less than λ_M; this is because the costs of failure are generally higher for the user than for the manufacturer.

The user has to make two decisions:

(a) To purchase an item of equipment for which TLOC is a **minimum**.
(b) To choose a **maintenance strategy** for the item which again ensures that TLOC is a minimum.

In order that these decisions can be made correctly, a full analysis of the cost of failures and maintenance over the lifetime of the item/system must be carried out.

The simplest maintenance strategy is **breakdown maintenance**, also referred to as unscheduled or repair maintenance. Here the equipment is simply repaired or replaced when it fails; no other maintenance costs are incurred. This corresponds to the everyday situation of replacing or repairing a car tyre only when it punctures.

We can now develop an algebraic expression for the cost of failures and maintenance for a breakdown maintenance strategy on a given piece of equipment or system. If the system lifetime is T years and the average failure rate λ failures per year, then the total number of failures is λT. If the mean down time following failure (Sections 2.3.2 and 6.3.1) is MDT hours, then the total down time due to failure is $\lambda T(\text{MDT})$ hours. The total lifetime cost of failures is the sum of the repair cost (materials and labour) and the 'process' cost, *i.e.* the cost of lost production and efficiency while the system is withdrawn for repair. If we define:

£C_R = average materials cost per repair
£C_L = repair labour cost per hour
£C_{PB} = process cost per hour following a breakdown

then the total repair cost is $[C_R\lambda T + C_L\lambda T(\text{MDT})]$ and the total process cost is $C_{PB}\lambda T(\text{MDT})$ giving:

$$\text{Total lifetime cost of failures with breakdown maintenance} = [C_R + (C_L + C_{PB})(\text{MDT})]\lambda T \qquad (6.2)$$

We note that the repair labour and process cost terms involve the product λ MDT of mean failure rate and mean down time. From Equation 2.9, **unavailability** is given by:

$$U = \frac{\text{MDT}}{\text{MTBF} + \text{MDT}}$$

Since $\text{MTBF} = \dfrac{1}{\lambda}$

then:

$$U = \frac{\text{MDT}}{\dfrac{1}{\lambda} + \text{MDT}} = \frac{\lambda\text{MDT}}{1 + \lambda\text{MDT}}$$

Usually $\lambda\text{MDT} \ll 1$, giving $U \approx \lambda\text{MDT}$

These terms can therefore be also expressed in terms of unavailability U.

Once the system has been purchased, there is nothing the user can do to reduce λ if only a breakdown maintenance strategy is used. Similarly the degree of freedom for reducing costs C_R, C_L and C_{PB} is probably limited. The main approach to minimizing the total cost of failures is to minimize the mean down time, MDT. Having a low MDT is one method of compensating for a high failure rate λ. An analysis of the different activities which contribute to MDT is carried out in Section 6.3.1.

The second maintenance strategy is **preventive maintenance** or scheduled maintenance. Here equipment is serviced and/or components replaced at regular fixed intervals. We can define the **maintenance** or **service interval** T_M years as the time between routine services. In the simple example of a car tyre, this corresponds to replacing the tyre either at fixed intervals of time, *e.g.* two years, or at fixed mileage intervals, *e.g.* 25 000 miles. The theory is that this will reduce excessive tyre wear and the probability of a puncture. The problem here is that tyre wear depends on many factors and that some tyres may wear out in less than 25 000 miles and others are only slightly worn at 25 000 miles.

We can again develop an algebraic expression for the cost of a preventive maintenance strategy. The **maintenance frequency** m, *i.e.* the number of times per year routine servicing is carried out, is equal to $1/T_M$. The total number of routine services is mT and if MMT hours is the **mean maintenance time**, then the total down time due to routine maintenance is $mT(MMT)$ hours. If we define:

$\pounds C_M$ = average materials cost per service
$\pounds C_{PP}$ = process cost per hour during servicing

then the total servicing cost is $[C_M mT + C_L mT(MMT)]$ and the total process cost is $C_{PP} mT$ (MMT) giving:

$$\text{Total lifetime cost of preventive maintenance} = [C_M + (C_L + C_{PP})(MMT)]mT \tag{6.3}$$

However, it is not practicable to carry out a preventive maintenance strategy only; equipment must be repaired if it breaks down. In many industrial situations a policy of both breakdown and preventive maintenance strategies is pursued. From Equations 6.2 and 6.3 we have:

$$\text{Total lifetime cost of breakdown and preventive maintenance} = \begin{array}{l}[C_R + (C_L + C_{PB})(MDT)]\lambda T \\ + [C_M + (C_L + C_{PP})(MMT)]mT\end{array} \tag{6.4}$$

and we see that preventive maintenance has introduced another cost term.

Example 6.1

In Example 5.1 we saw that a manufacturer markets six different types of transmitter at quality levels 1 to 6. Use the data given below to decide:

(a) which transmitter the user should buy;
(b) whether the user should follow a breakdown and preventive maintenance strategy or a breakdown only strategy, once the selected transmitter is out of the guarantee period. The preventive maintenance strategy is that recommended by the manufacturer.

Data

Initial cost C_I = Manufacturer's selling price per unit s (see table)
Failure rate with preventive maintenance λ_{PM} (see table)
Failure rate without preventive maintenance $\lambda_{BM} = 0.1$

Average materials cost per repair $C_R = £20$
Repair/maintenance cost per hour $C_L = £15$
Process cost per hour following a breakdown $C_{PB} = £1000$
Average materials cost per service $C_M = £20$
Process cost per hour during servicing $C_{PP} = 0$ (process shut down)
Mean down time during breakdown MDT = 8 hours
Mean maintenance time MMT = 8 hours
Maintenance frequency $m = 1.0$ per year (manufacturer recommended)
System lifetime $T = 10$ years.

Quality level	$C_I = s£$	λ_{PM} yr^{-1}	TLOC £
1	320	0.5	42 420
2	450	0.2	18 130
3	700	0.1	10 240
4	1100	0.05	6 570
5	1400	0.02	4 428
6	2000	0.01	4 214

Solutions

(a) $\quad$ TLOC $= C_I + [C_R + (C_L + C_{PB})\text{MDT}]\lambda T + [C_M + (C_L + C_{PP})\text{MMT}]mT$
$\quad\quad\quad\quad = C_I + [20 + (15 + 1000)8]\lambda 10 + [20 + (15 + 0)8]m10$
$\quad\quad\quad\quad = C_I + 81\ 400\lambda + 1400m$

The table shows values of TLOC in **bold**. We see that TLOC is minimum for the transmitter with quality level 6 and the user should buy this transmitter. We note that the manufacturer's profit was maximum at quality level 4.

(b) $\quad$ For the selected transmitter, TLOC with breakdown and preventive maintenance is **£4214**. For a breakdown only strategy we have $\lambda = \lambda_{BM} = 0.1$ and $m = 0$. We now have:

$\quad\quad$ TLOC $= C_I + 81\ 400\lambda + 1400m$
$\quad\quad\quad\quad\quad = 2000 + 81\ 400 \cdot 0.1 + 0$
$\quad\quad$ **TLOC = £10 140**

Therefore a breakdown and preventive maintenance strategy is better.

6.2.2 Choice of maintenance strategy

Preventive maintenance is only economically viable if the total cost of breakdown plus preventive maintenance is less than the total cost of breakdown maintenance only. From Equations 6.2 and 6.4, the condition for preventive maintenance to be financially worthwhile is:

$$\begin{aligned} &[C_R + (C_L + C_{PB})(MDT)]\lambda_{PM}T \\ &+ [C_M + (C_L + C_{PP})(MMT)]mT \\ &< [C_R + (C_L + C_{PB})(MDT)]\lambda_{BM}T \end{aligned} \qquad (6.5)$$

In the above condition λ_{PM} denotes failure rate with preventive maintenance and λ_{BM} failure rate with breakdown maintenance only; this allows for the possibility that preventive maintenance causes a reduction in failure rate. Condition 6.5 will be satisfied in two general situations:

(a) **Preventive maintenance causes a significant reduction in failure rate**
In a situation where materials and process costs are comparable, i.e. $C_R \approx C_M$, $C_{PB} \approx C_{PP}$, and mean down time is also comparable with mean maintenance time, i.e. MDT $\approx$ MMT, then the magnitude of the square bracketed terms are comparable and condition 6.5 simplifies approximately to:

$$\lambda_{PM} + m < \lambda_{BM} \quad i.e. \quad \lambda_{PM} < \lambda_{BM} - m \qquad (6.6)$$

This means that, for preventive maintenance to be financially viable in this situation, mean failure rate with preventive maintenance must be less than the breakdown only failure rate minus the frequency of routine servicing.

(b) **Cost of a breakdown failure is significantly greater than cost of a routine service** Situations can occur when the cost of a breakdown failure is significantly greater than the cost of a routine service, i.e.:

$$C_R + (C_L + C_{PB})MDT \gg C_M + (C_L + C_{PP})MMT \qquad (6.7)$$

Assuming the materials cost of a breakdown repair is comparable with the materials cost of a routine service, i.e. $C_R \approx C_M$, condition 6.7 simplifies approximately to:

$$(C_L + C_{PB})MDT \gg (C_L + C_{PP})MMT \qquad (6.8)$$

Condition 6.8 is satisfied if **one or both** of two situations occur:

(i) MDT $\gg$ MMT

Mean down time is much greater than mean maintenance time. An analysis of the different activities which contribute to MDT is carried out in Section 6.3.1, but sudden failure will mean that MDT will include times for activities such as:

> Fault location and diagnosis
> Assembly of necessary repair personnel
> Assembly of necessary repair equipment

which will not be present in MMT. Fault location and diagnosis times can be long in complex equipment and systems. For heavy process equipment, such as pumps or compressors, the time required to assemble the necessary repair equipment (*e.g.* cranes) and skilled personnel may be many days.

(ii) $C_{PB} \gg C_{PP}$

Process cost per hour following an unplanned breakdown is much greater than the process cost per hour during routine servicing. If the process is shut down in an unplanned way, significant waste may occur due to loss of raw materials or product, production of off-specification materials, energy losses, *etc*. With a sudden shut-down it may be impossible to supply customers from alternative production facilities or from storage. In both cases C_{PB} could be very large. However, routine maintenance can take place at a time when the process is planned to be either shut down or operating at reduced capacity and alternative arrangements to supply customers have been made. In these situations C_{PP} may be very small or negligible.

If either condition (i) and/or condition (ii) is satisfied then condition 6.7 is satisfied. In condition 6.5, for preventive maintenance to be financially worthwhile, the preventive maintenance cost term can be neglected giving the approximate condition:

$$[C_R + (C_L + C_{PB})(MDT)]\lambda_{PM}T < [C_R + (C_L + C_{PB})(MDT)]\lambda_{BM}T$$

i.e.

$$\lambda_{PM} < \lambda_{BM} \tag{6.9}$$

Thus from conditions 6.6 and 6.9, we see that in both situations (a) and (b) the mean failure rate with preventive maintenance must be less than the failure rate with breakdown maintenance, for preventive maintenance to be financially worthwhile. If condition 6.5 is not met, then preventive maintenance is not economically viable.

The third maintenance strategy is **on-condition maintenance**. Here the condition of the equipment is monitored or observed and when it shows signs of wear, or some other indication that failure might be imminent, it is repaired or replaced. The process of condition monitoring should ideally show which element or component in the equipment or system is likely to fail and should be repaired or replaced. Returning to the example of the car tyre, on-condition maintenance corresponds to replacing or repairing the tyre when it shows signs of significant wear or damage. On-condition maintenance is similar to breakdown maintenance except that repair/replacement takes place **before** rather than **after** failure. The economic advantages of on-condition maintenance over breakdown maintenance can be explained by again examining the breakdown maintenance cost term

$$[C_R + (C_L + C_{PB})(MDT)]\lambda_{BM}T$$

The mean failure rate with on-condition maintenance will be exactly the same as the mean breakdown maintenance failure rate λ_{BM}. However, with on-condition maintenance there will be an interval of time, termed the **potential failure interval**, between the condition of the equipment showing that a given failure is imminent and that failure actually occurring (Section 6.5). Assuming that condition monitoring has correctly identified the fault, then during the potential failure interval the equipment and personnel necessary to repair the fault can be assembled and the process shut down in an orderly way. This means that the mean down time, MDT, with on-condition maintenance is much less than with breakdown maintenance. Furthermore, if the potential failure interval can be used to make alternative arrangements to supply customers while the equipment is down for repair, then this together with an orderly shut-down will mean that the process cost C_{PB} with on-condition maintenance will be much less than with breakdown maintenance. Thus on-condition maintenance is economically preferable to breakdown maintenance.

We can now compare the cost of an on-condition maintenance strategy with that of breakdown plus preventive maintenance. Because on-condition maintenance significantly reduces the cost of a breakdown failure, the only situation where breakdown plus preventive maintenance is likely to be financially better than on-condition maintenance is when it produces a significant reduction in failure rate (situation (a) and condition 6.6). However, in any detailed cost comparison, the cost of the monitoring equipment plus any other monitoring costs must be included in the total lifetime cost equation.

The above economic analysis also provides data on the size of the maintenance effort required for a given system. Thus, for the example of breakdown and preventive maintenance, Equation 6.4 tells us that on average $(MDT)\lambda + (MMT)m$ hours of maintenance effort are required on the system in each year. The provision of spare parts/components must be compatible with an average failure rate of λ per year and a service frequency of m per year for the given system. The total annual maintenance cost for the system is $[C_R + C_L(MDT)]\lambda + [C_M + [C_M + C_L(MMT)]m$.

6.3 Breakdown maintenance

6.3.1 Factors contributing to mean down time MDT

As discussed above, a breakdown maintenance strategy simply involves repairing or replacing equipment when it fails. The previous section also showed that the main approach to minimizing the total cost of failures with a breakdown maintenance only strategy is to minimize the mean down time MDT. Down time was defined in Section 2.3.2 as the time interval that the equipment is in a failed state; since the down time for a given piece of equipment is not constant, but varies randomly according to circumstances, it is necessary to define a mean down time MDT. Figure 6.2 shows the main activities that may have to take place while the equipment is 'down'; each activity has a time interval associated with it which contributes to the down time.[1] These time elements are:

(a) **Realization time** This is the time which elapses before the fault becomes apparent. This can be minimized by the provision of relevant alarms.

(b) **Access time** This is the time that elapses between the realization that a fault exists and the commencement of fault finding. It includes the time involved in isolating the equipment and making it safe, the removal of covers and shields, the connection of test equipment. It can be minimized by good mechanical design, *i.e.* ensuring that elements with the highest failure rate are the most accessible.

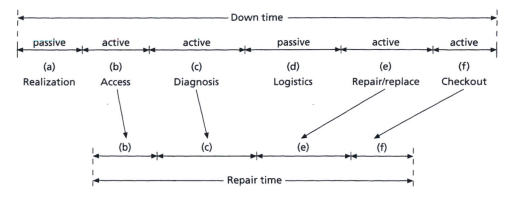

Figure 6.2   Down time and repair time

(c)   **Diagnosis time** This is the time taken to diagnose the fault and includes the collection and analysis of data, use of diagrams and algorithms. An explanation of fault diagnosis techniques is given in the following section.

(d)   **Logistics time** This is the time taken to assemble the necessary spare parts/components, other equipment and personnel necessary to repair the fault. It is minimized by having an adequate spares holding and a pool of repair equipment/personnel on-site.

(e)   **Repair/replacement time** This is the time taken to repair the fault or replace the component. It is again minimized by good mechanical design which allows easy access.

(f)   **Checkout time** This is the time taken to verify that the fault condition no longer exists.

Activities (b), (c), (e), (f) are called active repair elements and (d) a passive repair element. Repair time is the sum of the active repair elements, *i.e.* (b) + (c) + (e) + (f) (Figure 6.2). This again will vary randomly and it is again necessary to define a mean value which is normally referred to as **mean time to repair** (MTTR). The use of MTTR rather than MDT can cause confusion; since MTTR only involves active repair elements it may be significantly less than MDT.

6.3.2 Fault diagnosis

In the previous section we saw that the diagnosis time (c) could be a significant element of MDT. In order to minimize diagnosis time, a systematic method of fault diagnosis is necessary, based on the collection and analysis of relevant data. In many situations the fault is in an element or component of a system which is itself an element of a larger system and so on. Figure 6.3 shows a hierarchy of four systems: the overall plant is the highest level, then the plant item, then the control loop, then the element at the lowest level. Each system is represented by a **functional block diagram**; here each block represents a clearly identifiable function, *e.g.* a pump, an amplifier or a controller. Figure 6.3

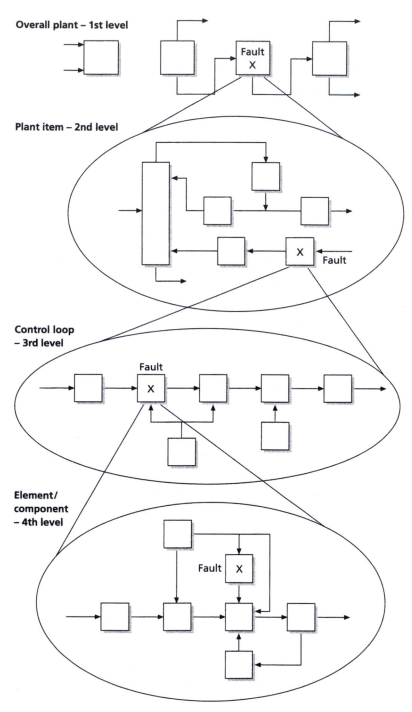

Figure 6.3 Hierarchy of functional block diagrams

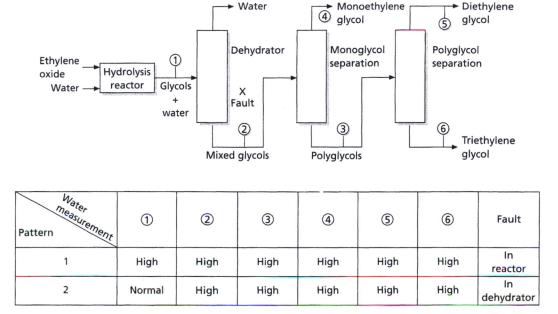

Figure 6.4 Functional block diagram and fault diagnosis for overall plant

shows the overall fault diagnosis strategy. Study of the overall plant shows that the fault lies in a given plant item, study of the plant item shows that the fault lies in a given control loop, study of the control loop shows that the fault lies in a given element and finally, study of the element shows that the fault lies in a given component. Thus the diagnosis strategy is to move down the hierarchy, gradually narrowing down the source of the fault. This logical process is now illustrated by an example of the location of a fault in a chemical plant.

Figure 6.4 is a functional diagram of an overall plant to produce mono-, di- and tri-ethylene glycol from an ethylene oxide feedstock. The overall fault is the presence of too much water in the three glycol products. Measurements of water content are taken at points 1 to 6 and from the pattern of these measurements it is possible to locate the position of the fault. For example, if measurements 1 to 6 are all high (Pattern 1), then the fault lies in the reactor (or upstream of the reactor); if measurement 1 is normal and 2 to 6 are high (Pattern 2), then the fault lies in the dehydrator. We assume that pattern 2 is obtained so that the fault lies in the dehydrator.

Figure 6.5 is a functional block diagram of the glycol dehydrator, the fault is the presence of too much water in the glycol product stream. The diagram shows the eight measurements available on the plant to aid fault diagnosis, these are one temperature, two pressures, two levels and three flow rates. A large number of fault patterns are possible, each one corresponding to a given fault. Four patterns are shown: Pattern 1 corresponds to failure of the steam flow control system, Pattern 2 to failure of the steam supply, Pattern 3 to failure of the cooling water supply and Pattern 4 to failure of the reflux pump. We

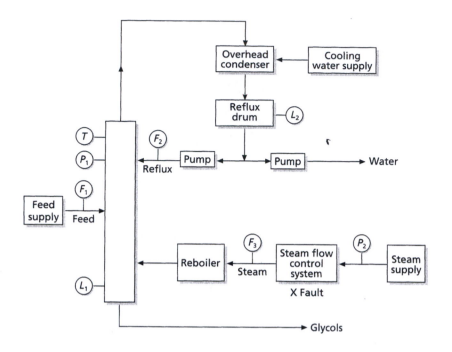

Measurement Pattern	Column temperature T	Column pressure P_1	Base level L_1	Reflux drum level L_2	Feed flow rate F_1	Reflux flow rate F_2	Steam flow rate F_3	Steam pressure P_2	Fault
1	Low	Low	High	Low	Normal	Normal	Low	Normal	Steam flow control system
2	Low	Low	High	Low	Normal	Normal	Low	Low	Steam supply
3	High	High	Low	Low	Normal	Low	Normal	Normal	Cooling water supply
4	High	High	Low	High	Normal	Low	Normal	Normal	Reflux pump

Figure 6.5   Functional block diagram and fault diagnosis for glycol dehydrator

assume that Pattern 1 is obtained, *i.e.* the steam flow control has failed, this is consistent with the presence of too much water in the glycol product.

Figure 6.6(a) shows the steam flow control system in conventional form, and Figure 6.6(b) shows the same system in functional block diagram form. The functional block diagram indicates the position of eight necessary test measurements:

steam flow rate, Q
orifice plate differential pressure, ΔP
power supply voltage, V
controller measured value current, i_1

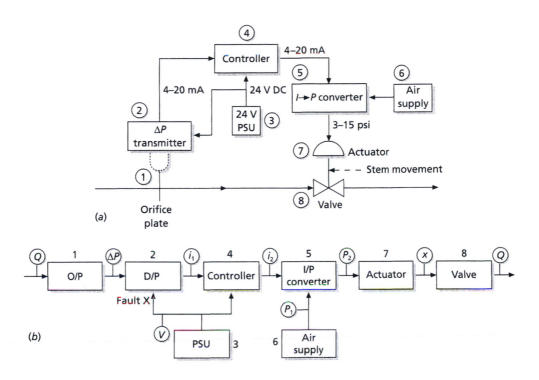

Figure 6.6 Fault diagnosis in steam flow control system:
(a) steam flow control system;
(b) functional block diagram;
(c) possible fault patterns

Pattern \ Measurement	Steam flow rate Q	Differential pressure ΔP	Supply voltage V	Controller measured valve i_1	Controller output i_2	Air supply pressure P_1	Converter output pressure P_2	Actuator position x	Fault
1	Low	Low	Normal	High	Low	Normal	Low	Low	D/P transmitter
2	Low	Low	Low	Low	Low	Normal	Low	Low	Power supply unit
3	Low	Low	Normal	Low	High	Low	Low	Low	Air supply
4	High	High	Normal	High	High	Normal	High	High	Controller
5	Low	High	Normal	High	Low	Normal	Low	Low	Orifice plate blockage

(c)

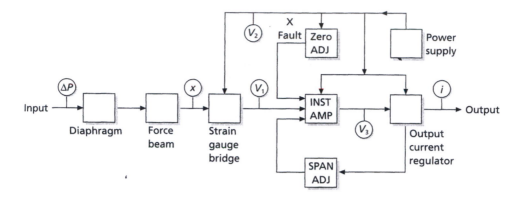

Measurement ⟍ Pattern		Input differential Pressure ΔP Pa	Force beam position x mm	Strain gauge O/P V_1 mV	Power supply output V_2 V	Inst amplifier output V_3 V	Output current i mA	Fault
Normal correct	Min	0	0	0	12.0	1.0	4.0	
pattern	Max	3×10^4	1.0	10.0	12.0	5.0	20.0	
Fault pattern	Min	Normal	Normal	Normal	Normal	1.50	6.0	Zero adjustment
1	Max					5.50	22.0	
Fault pattern	Min	Normal	Normal	Normal	Normal	1.0	4.0	Span adjustment
2	Max					5.25	21.0	
Fault pattern	Min	Normal	0	0	Normal	1.0	4.0	Diaphragm/ force beam
3	Max		0	0		1.0	4.0	
Fault pattern	Min	Normal	Normal	Normal	Normal	1.0	4.0	Instrument amplifier
4	Max					1.0	4.0	

Figure 6.7 Functional block diagram and test data sheet for differential pressure transmitter

controller output current, i_2
converter air supply pressure, P_1
converter output pressure, P_2
actuator position, x

Again a large number of fault patterns are possible; Figure 6.6(c) shows five such patterns each with the corresponding fault. We assume that Pattern 1 is obtained which corresponds to a fault in the differential pressure transmitter. Here Q is low causing ΔP to be correspondingly low; however, the D/P transmitter fault results in the D/P output current i_1 being high. This means that the controller measured value is high and above the set point value; this causes the controller to set output current i_2 low. This in turn causes

P_2, x and Q to be low. The fault is then that the differential pressure transmitter output current is too high.

Figure 6.7 shows the functional block diagram of the differential pressure transmitter; when working correctly the transmitter should give an output current in the range 4–20 mA proportional to input differential pressure in the range $0–3 \times 10^4$ Pa. The diagram indicates the position of six necessary test measurements:

input differential pressure, ΔP
force beam position, x
strain gauge bridge output voltage, V_1
power supply voltage, V_2
instrumentation amplifier output voltage, V_3
output current, i

The figure also shows a **test data sheet** for the transmitter. This shows first the minimum and maximum values that should be obtained at each test point, if the transmitter is working correctly. Again a large number of fault patterns are possible and the test data sheet shows four such patterns each with the corresponding fault. We assume that fault pattern 1 is obtained, which corresponds to a fault in the zero adjustment unit; closer inspection of this unit shows that a potentiometer has been incorrectly adjusted causing an output current of 6.0 mA at zero input ΔP instead of 4.0 mA.

Figure 6.8 shows an **algorithm** for fault location in the differential pressure transmitter; this is a flow chart of tests, and progress through the flow chart is determined by the answers to a series of questions to which the only possible answers are 'Yes' and 'No'. The use of algorithms may reduce the diagnosis time because questions relating to more likely faults are asked first.

6.4 | Preventive maintenance

In Section 6.2.1 we defined preventive maintenance as the servicing of equipment or replacement of components at regular fixed intervals T_M, where T_M is the **maintenance** or **service interval**. For preventive maintenance to be economically worthwhile the total cost of breakdown plus preventive maintenance must be less than the cost of breakdown maintenance alone (condition 6.5). We saw that this condition is satisfied in two situations; in both cases the failure rate with preventive maintenance must be less than the failure rate with breakdown maintenance (conditions 6.6 and 6.9). These conditions are likely to be met when preventive maintenance stops the equipment entering a wear-out failure region (Section 2.6). In Section 5.5.5 the Pareto distribution of failure rate was discussed. This is based on reliability data obtained from field tests on a given system and shows the frequency of failure for the different components in the system. It highlights the few (normally) critical components with high failure rate and is therefore vital in the design effort to improve reliability. It can also be used to devise preventive maintenance strategies. By measuring the Pareto distribution at regular intervals throughout the lifetime of the system, it is also possible to identify those components

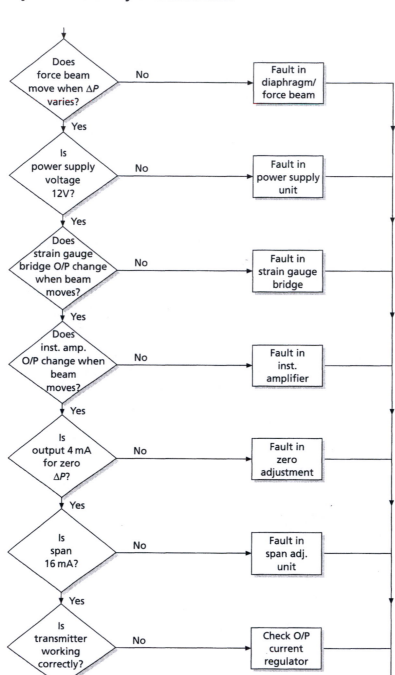

Figure 6.8 Algorithm for fault location in differential pressure transmitter

whose failure rate increases with time. These are exhibiting wear-out failure and will benefit from preventive maintenance. This section shows how **preventive replacement** and **routine preventive maintenance** can be used to prevent wear-out failure. The use of **routine testing** to check for unrevealed or dormant faults in protective or standby systems is then discussed. Finally there is a brief analysis of the factors that influence **mean maintenance time** (MMT).

6.4.1 Preventive replacement

Many mechanical/hydraulic components (*e.g.* oil filters, clutch mechanisms and diaphragms) and some electrical/electronic components (*e.g.* electrolytic capacitors, light bulbs and microwave valves) are characterized by **wear-out failure** (Section 2.6). Figure 6.9 shows an idealized form of the variation of failure (hazard) rate λ with time t for such components; λ has a constant value λ_0 up to time $t = \bar{T}_W$ and then increases rapidly for $t > \bar{T}_W$. $\bar{T}_W$ is termed the **mean wear-out time** for the components. The figure shows that wear-out times for individual components are distributed randomly about $\bar{T}_W$; we assume a normal or Gaussian distribution with standard deviation σ. If the component lifetime $\bar{T}_W$ is significantly less than the system lifetime T then replacement of components before wear-out occurs is economically worthwhile. If the replacement time T_M is set at two standard deviations before $\bar{T}_W$, *i.e.*

$$T_M = \bar{T}_W - 2\sigma \tag{6.10}$$

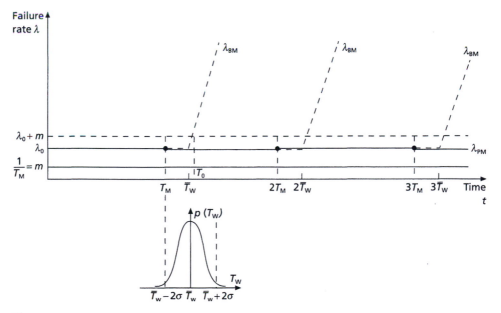

Figure 6.9 Preventive replacement strategy

then there is a 2.3% chance that a component will fail before replacement (Table 1.2). If T_M is three standard deviations before $\bar{T}_W$, *i.e.*

$$T_M = \bar{T}_W - 3\sigma \tag{6.11}$$

then the probability of failure before replacement is reduced to 0.1%. The corresponding replacement frequency m is equal to $1/T_M$. Figure 6.9 shows how this stategy prevents the component from entering the wear-out failure region so that the failure rate with preventive maintenance is held at λ_0, *i.e.* $\lambda_{PM} = \lambda_0$. The figure also shows that condition 6.6, *i.e.* that $\lambda_{PM} + m$ should be less than λ_{BM} (the failure rate with breakdown maintenance), is satisfied for $t > T_0$.

6.4.2 Routine preventive maintenance

This is the routine maintenance of components, especially mechanical and hydraulic components, in order to increase mean wear-out time $\bar{T}_W$ and again prevent the component entering the wear-out failure region. This involves such activities as changing lubricating oil, changing hydraulic fluid, greasing of bearings, adjustment of clearances, cleaning and painting to reduce corrosion. Figure 6.10 shows an ideal situation; here routine maintenance on a given component is carried out before the wear-out time $\bar{T}_W$ and results in an extension of wear-out time by an amount ΔT_W. Further routine maintenance should be carried out before $\bar{T}_W + \Delta T_W$ and again before $\bar{T}_W + 2\Delta T_W$, *etc.*; the maintenance interval T_M should therefore equal ΔT_W (and maintenance frequency $m = 1/\Delta T_W$). This activity will not always result in an improvement in wear-out times and a comparison between wear-out times with and without preventive maintenance should be carried out initially.

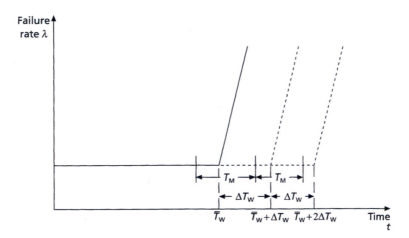

Figure 6.10   Routine preventive maintenance

6.4.3 Routine testing for unrevealed faults

Protective systems with majority voting were discussed in Section 3.5 and standby systems in Section 3.6; these systems are only required to operate in emergency situations. Some faults in the system may only become apparent when the system fails to operate during the emergency situation and are termed *unrevealed* or *dormant* faults. Unrevealed faults may prevent the system operating when it should and therefore may result in a **fail-danger failure** (Section 7.5.1). To avoid this serious situation the system is regularly and periodically tested. The **testing interval** T_M is the time between the successive tests on the system and **mean time to test** (MTTT) is the mean time to carry out the test. Assuming that the system failure rate associated with the unrevealed fault has a constant value λ, then the probability of a system failure at any time t is given by the exponential relation (Equation 2.16):

$$F(t) = 1 - e^{-\lambda t} \tag{6.12}$$

Just after the system has been tested and proved to be working correctly, the unreliability F can be assumed to have been restored to zero, *i.e.* $F = 0$. Thereafter, the unreliability increases exponentially to the maximum value $F_{MAX} = 1 - e^{-\lambda T_M}$ just before the next test. The time variation of F therefore has the saw tooth waveform shown in Figure 6.11. A value for maximum unreliability F_{MAX} should be specified and this enables the required testing interval T_M to be found. Thus if $F_{MAX} = 10^{-3}$ and $\lambda = 10^{-2}$ faults/year we have:

$$10^{-3} = 1 - \exp(-10^{-2}T_M) \tag{6.13}$$

giving $T_M = 0.1$ years, *i.e.* a test frequency m of approximately once per month. The above simple analysis assumes that MTTT is much smaller than T_M. If this is not the case and MTTT = 1 day, for example, then the system unavailability due to testing is 12 days per year or 3.3×10^{-2}. This is significantly larger than the unavailability due to failure (fractional dead time) which is $\frac{1}{2}F_{MAX}$, *i.e.* 0.5×10^{-3}. The influence of testing interval on the fail-danger probability and fractional dead time of protective systems is further discussed in Chapter 7.

Unrevealed or dormant failures can also occur in systems which are always operational. For example, the failure of a capacitor in an amplifier circuit may affect the frequency response of the circuit without affecting the gain, so that while some degradation of the output signal occurs, it is insufficient to be registered as a failure.

6.4.4 Factors contributing to mean maintenance time MMT

In the three previous sections we saw how the maintenance interval T_M or maintenance frequency $m = 1/T_M$ is determined in three different forms of preventive maintenance activity. The analysis of Section 6.2.1 showed that the cost of preventive maintenance is also critically dependent on the mean maintenance time MMT (Equation 6.3). In Section 6.3.1 we analysed the activities that contribute to mean down time MDT in a

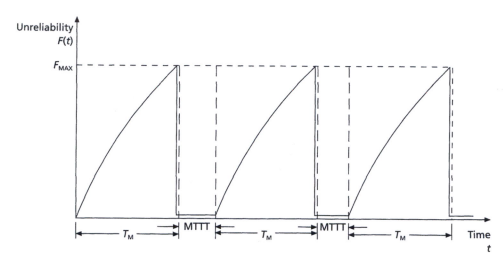

Figure 6.11   Routine testing for unrevealed faults

breakdown maintenance activity (Figure 6.2). Some of these activities will be present in preventive maintenance and will contribute to MMT and some will not. Since there is no fault in preventive maintenance, the time elements associated with Realization (a) and Fault diagnosis (c) will not be present. Similarly, since the materials, equipment and personnel necessary can be assembled before the system is shut down for maintenance, the Logistics time (d) should not be present. However, preventive maintenance will involve Access (b), Servicing/replacement (e) and Checkout (f) activities so that:

$$\underset{(b)}{\textbf{maintenance time} = \textbf{access time}} + \underset{(e)}{\textbf{servicing/replacement time}}$$
$$\underset{(f)}{+ \textbf{ checkout time}} \tag{6.14}$$

Access time can be minimized by good mechanical design so that components/elements that are replaced/serviced most frequently are the most accessible. Also components/elements that are accessed via the same route are, where possible, replaced/serviced simultaneously. For a given item of equipment, all three activities in Equation 6.14 will vary randomly producing corresponding random variations in maintenance time; it is therefore necessary to define **mean maintenance time** MMT. In general MMT will be shorter than **mean time to repair** MTTR, which is the sum of the four active repair elements (b), (c), (e) and (f) in breakdown maintenance (Figure 6.2). Also both MMT and MTTR are shorter than MDT which involves all six elements (a) to (f) in breakdown maintenance. Table 6.1 gives data for the manhours required for the maintenance of instruments in the process industries.[2] The table gives mean, range and standard deviation for each type of instrument, but it is not clear whether the data relate to breakdown or preventive maintenance or a combination of both.

Table 6.1 Manhours required for the maintenance of instruments in the process industries (after Lees[2])

Instrument	Maintenance manhours		
	Mean (h/year)	Range (h/year)	Standard deviation (h/year)
Control valve			
Globe	5	4–6	0.41
Butterfly	7	6–9	0.76
Saunders	7	6–8	0.70
Pressure measurement			
Pressure transducer			
Bourdon tube	4	3–5	0.41
Bellows	3	2–5	0.70
Flow measurement			
Differential pressure transducer			
Diaphragm	6	4–8	0.70
Bellows	6	5–7	0.70
Mercury	8	6–10	0.26
Magnetic flowmeter	8	6–10	0.26
Turbine meter	10	8–12	0.24
Positive displacement meter	10	8–12	0.24
Level measurement			
Differential pressure transducer			
Diaphragm	5	4–7	0.81
Bellows	4	3–6	0.76
Mercury	5	4–6	0.35
Float	5	4–6	0.35
Capacitance	6	4–8	0.66
Displacer	5	4–7	0.66
Radiation	10	8–12	2.1
Bubble pipe	4	3–5	0.35
Temperature measurement			
Fluid bulb			
Force balance	4	3–6	0.46
Motion balance	4	3–8	0.44
Controller			
Pressure	5	4–7	2.5
Flow	6	5–8	0.82
Level	5	4–7	2.5
Temperature	5	4–7	2.5
Analyser			
pH	29	20–40	6.21
Gas chromatograph	145	130–170	2.5
O_2	38	35–42	2.1
CO_2	21	18–25	2.1

Table 6.1 Cont'd

Instrument	Maintenance manhours		
	Mean (h/year)	Range (h/year)	Standard deviation (h/year)
Infrared	76	62–94	5.69
Boiling point	80	74–85	4.82
Combustibles	50	44–61	3.74
Density			
Displacement	34	29–40	3.16
Radiation	40	33–49	4.00
Recorder or indicator			
Pressure	4	3–6	0.58
Flow	6	5–8	0.49
Level	5	4–7	0.70
Temperature	4	3–6	0.58

6.5 | On-condition maintenance

As discussed in Section 6.2.2, an **on-condition maintenance** strategy involves monitoring some condition or variable associated with the equipment and then replacing/repairing the equipment when that condition shows that failure may be imminent. For this strategy to work, there must be a measurable condition or variable, associated with a given failure mode in a given item of equipment, which decreases (or increases) continuously and monotonically with time as that failure is approached. Figure 6.12 shows a condition or variable decreasing with time as failure is approached; the curve shows that there are two important values of the condition. The first value corresponds to there being a significant probability of imminent failure, *i.e.* to **potential failure**. The second value corresponds to the equipment being unable to perform its normal function, *i.e.* to **functional failure**. The time interval between potential failure and functional failure is termed the **potential failure interval** PFI. It is important that the equipment user establishes this **condition versus time** variation for a given failure in a given type of equipment. At the very least the condition values corresponding to potential and functional failure and the potential failure interval should be known. However, the user will only be able to establish a mean curve and mean parameters for a given type of equipment; values for individual items of equipment will be distributed statistically about the means.

The simplest method of condition monitoring involves the four human senses of **sight**, **sound**, **touch** and **smell**. Thus sight can be used to detect a worn brake pad or leaking valve, sound to detect an exhaust system failure, touch or smell to detect an overheated electric motor. However, such information will be **qualitative**, *e.g.* a motor will feel 'cold', 'warm' or 'hot', and **subjective**, *i.e.* will depend on the senses, skill, judgement and commitment of the human observer.

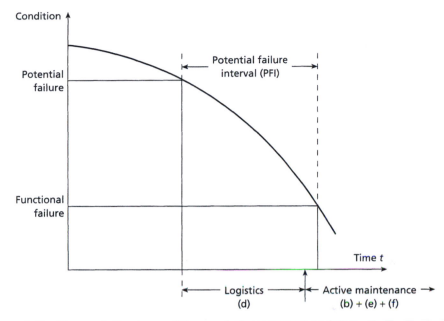

Figure 6.12   Time variation in condition on during approach to failure

Several basic measurements provide quantititative, objective information on equipment condition. **Temperature** measurements will provide information on the conditions of bearings, furnaces, lubricating oil, boiler tube metal and electric motor windings. **Strain** measurements will provide information on the condition of pressure vessel concrete and boiler liners. **Vibration** measurements, *e.g.* displacement, velocity or acceleration amplitudes, will provide information on the condition of rotating machinery such as pumps and compressors. **Pressure** measurements will provide information on the condition of fluid distribution systems, *e.g.* oil supply systems; **differential pressure** provides information on the condition of filters, distillation columns and pumps. **Liquid level** measurements provide information on the loss or ingress of liquids in distribution systems. **Displacement** measurements provide information on machine movement, for example the 'float' in pump shafts or rack and pinion systems. **Electric current** measurements provide information on the performance of electric motors and pumps.

Vision systems can provide detailed information on the condition either of the surface of equipment or within the bulk material.

Optical vision systems can provide information on surface condition. These use line scan cameras remote from the equipment which are based on arrays of charge-coupled devices (CCD). A typical array will consist of many thousands of elements arranged in a rectangular matrix. Each element consists of a silicon photodiode with an MOS (metal–oxide–silicon) capacitor. The photodiodes respond to light with wavelengths between approximately 0.45 and 0.9 μm, *i.e.* in the visible and near infrared: the charge on the capacitance depends on the intensity of illumination of the target surface. The diodes in the array are sequentially scanned at a high pulse rate, the height of each pulse being proportional to the light intensity falling on the element. This pulse train is input to a visual display unit (VDU) which displays the condition of the surface. In hostile environments involving extremes of heat, voltage, radiation or corrosion, or where there is limited space or access, **optical fibres** can be used to transmit optical information to a remote camera or receiver. **Eddy current** techniques can also be used to monitor both the surface of metal objects and the condition up to 6 mm below the surface. Here a primary AC voltage is applied to the component via an excitation coil. This induces eddy currents in the component which in turn produce a secondary voltage in a pick-up coil. The relation between primary and secondary voltages gives an indication of surface condition.

There are two main types of vision systems suitable for monitoring the condition inside bulk material. **Radiographic** systems use either an X-ray or a gamma-ray source on one side of the specimen and a detector on the other. The amount of radiation received by the detector depends on the density at the test location; a crack causes a reduction in density and an increase in received radiation. In **ultrasonic** systems a pizeoelectric crystal acting as both transmitter and receiver is mounted on the component. A pulse of high frequency sound is transmitted into the component; if this encounters a crack then some of the sound energy is reflected back towards the crystal. The position of the crack can be found from 'the time of flight' of the returning pulse.

6.5.1 Periodic condition monitoring

Here the condition of the system is monitored at regular periodic intervals. The choice of **condition monitoring interval** T_c is determined by the potential failure interval PFI. Figure 6.13(a) shows the situation where the condition is monitored at intervals T_c, T_c being greater than PFI. The condition is satisfactory at monitoring instant 1, and also at monitoring instant 2. However, monitoring instant 3 occurs after functional failure is likely to occur so that a decision to replace/repair the equipment should be made at monitoring instant 2. Instant 2 is well in advance of the functional failure time so that some of the useful life of the equipment is lost. Figure 6.13(b) shows the situation when T_c is less than PFI. The condition is satisfactory at monitoring instants 1, 2 and 3 and unsatisfactory at instants 4, 5, 6 and 7. At instants 4 and 5 a decision can be made to either replace/repair the equipment or keep it running. This decision should be based on a comparison of the potential financial loss incurred by not using the full lifetime of the equipment and the potential financial loss incurred by a sudden failure which necessitates unplanned breakdown maintenance and a longer down time. The equipment would be repaired/replaced at instant 6, since instant 7 is after the functional failure time.

Thus in order to utilize as much of the equipment lifetime as possible, the condition monitoring interval should be significantly less than the potential failure interval. To minimize the cost of down time, the condition monitoring interval T_c should, where possible, coincide with the preventive maintenance interval T_M so that on-condition and preventive maintenance activities occur at the same times. Thus, in a modern family car both condition monitoring and preventive replacement/maintenance takes place at 6000 mile

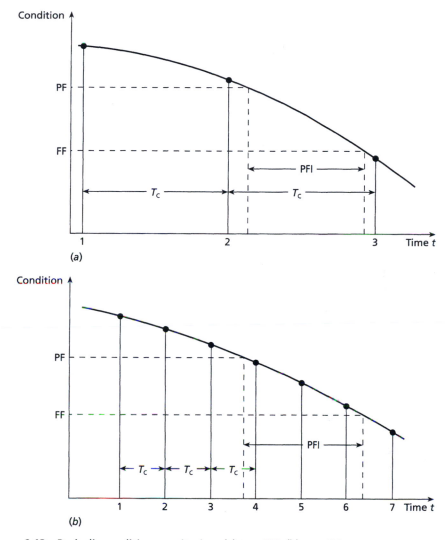

Figure 6.13 Periodic condition monitoring: (a) $T_c > $ PFI; (b) $T_c < $ PFI

intervals. For example, a 30 000 mile service includes 27 activities of which 18 are condition monitoring (checking tyres, brake pads, leaks, *etc.*) and 9 are preventive replacement/maintenance (changing oil filter, engine oil, brake fluid, *etc.*).

In Section 6.4.1 we saw that wear-out times for a given type of component can be assumed to follow a Gaussian distribution with mean value $\bar{T}_W$ and standard deviation σ. Thus, in a preventive maintenance strategy (Figure 6.9), **every** component must be replaced at a time T_M, which is 2σ to 3σ earlier than $\bar{T}_W$, in order to keep the probability of failure before replacement to an acceptably low value. This strategy means that with preventive maintenance a large number of items are repaired/replaced well before the end of their useful life. This is equivalent in Figure 6.13 to replacing **every** item at the potential failure time. Figure 6.13(b) shows that by monitoring the condition of an **individual** item, the replacement/repair of that item can be postponed until just before the functional failure time. As an example of the cost savings that can be achieved, consider a large pump that is returned to the manufacturer every six months for regular preventive maintenance at a cost of £25 000 each time. If, with condition monitoring, the interval between pump repairs can be increased on average to eight months then the average annual saving is £12 500.

6.5.2 Continuous condition monitoring

Here the condition of the equipment or system is monitored continuously; in general there are two situations where continuous monitoring is necessary.

(a) The potential failure interval PFI is small, *i.e.* only a few days or several hours. This means that the condition monitoring interval T_c would, in turn, have to be only a few hours, which is not very practicable. Moreover the entire PFI should be utilized to give the maximum time for the necessary repair equipment and labour to be assembled.

(b) The equipment or system is run continuously and the 'process' cost $£C_p$ of lost production during each hour the equipment is down is high. Condition monitoring systems should, therefore, be used which do not require the monitored system to be shut down. This may rule out some of the more complex monitoring systems, for example, ultrasonic and radiographic vision systems. In this situation it may be better to continuously monitor basic variables such as temperature and strain.

In Section 6.3.1 we saw that there were six elements which contributed to down time in a breakdown maintenance strategy (Figure 6.2). Some of these elements will not be present with on-condition maintenance. Since the system is continuously monitored the realization time element (a) should be negligible. Since the monitoring system has been designed so that ideally each potential failure mode is monitored by a given measured condition, the the fault diagnosis time element (c) should also be negligible. The logistics activity (d) involving the assembly of the necessary repair equipment and personnel should take place during the potential failure interval between potential and functional failure (Figure 6.12). The system ideally should then be shut down just before functional failure is likely to occur. The active maintenance activities of access (b), repair/replacement (e) and checkout (f) can then take place, so that only activities (b), (e) and (f) contribute to the mean down time with on-condition maintenance. Thus MDT with on-condition maintenance should be significantly less than with breakdown maintenance; this is provided that the logistics activity can be fitted into the potential failure interval.

Figure 6.14 shows a typical condition monitoring system; the input information consists of analogue measured variables and digital equipment states. There is a **sensing** element for each measured variable; this gives an output signal which depends in some way on the measured variable, *e.g.* a resistance or millivolt EMF. The output of each

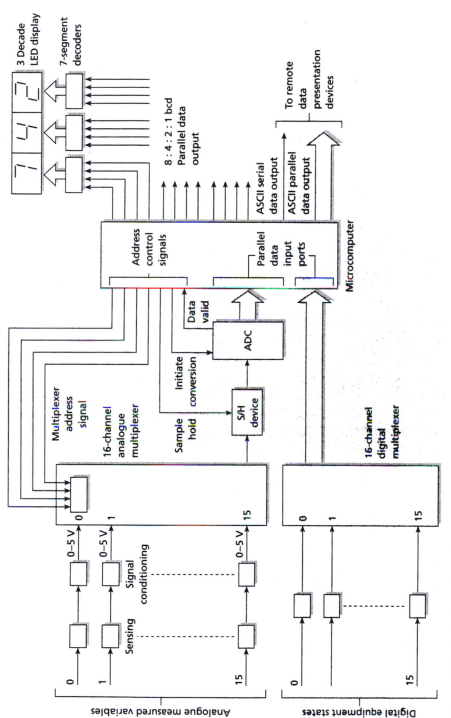

Figure 6.14 Typical condition monitoring system

Table 6.2 Typical sensing and signal conditioning elements for condition variables

Measured variable	Sensing element(s)	Signal conditioning elements
Temperature (contacting)	Thermocouple	Reference junction circuit + instrumentation amplifier
Temperature (non-contacting)	Optical system + infrared detector	Amplifier
Strain	Strain gauge	4 gauge bridge + amplifier
Level	Electronic differential pressure transmitter	Current (e.g. 4–20 mA) to voltage (e.g. 0–5 V) converter
Angular velocity	Optical tachogenerator	Frequency to voltage converter
Linear displacement	Linear variable differential transformer (LVDT)	AC amplifier + phase sensitive demodulator + low pass filter
Pressure (+ differential pressure)	Diaphragm + capacitance displacement sensor	AC bridge + AC amplifier + phase sensitive demodulator + low pass filter
Vibration { Acceleration	Piezoelectric crystal	Charge amplifier
Velocity	Piezoelectric crystal	charge amplifier + 1 stage of integration
Displacement	Piezoelectric crystal	Charge amplifier + 2 stages of integration
DC current	Standard resistor	Amplifier
AC current	Instrument transformer	Rectifier + low pass filter

sensing element is then input to a **signal conditioning** element which converts the sensor output to a common signal range, typically 0–5 V. Table 6.2 gives typical sensing and signal conditioning elements for measured variables used in condition monitoring. The voltage signals are then input to an **analogue multiplexer** which selects the signals either **sequentially** or **randomly** for further processing; the multiplexer shown has 16 input channels but this can be extended to 32 or 64. The selected signal is then input to a single **sample/hold** (S/H) device followed by a single **analogue-to-digital converter** (ADC); the S/H device follows the incoming signal but gives a constant output voltage during the time interval that the analogue-to-digital conversion takes place. The ADC gives a parallel digital output signal, typically 8, 10 or 12 bits, which passes to one of the parallel input interfaces of the microcomputer.

The digital equipment state inputs can take one of two values, typically 0 V for a logic '0' and 15 V for a logic '1'. These values correspond, for example, to a motor being switched on or off, a pump being tripped or not tripped, or a valve being closed or open. These digital signals are passed to a **digital multiplexer** which selects a group of typically

8, 16 or 32 either sequentially or randomly. The selected group forms a parallel digital signal which again passes to one of the parallel input ports of the microcomputer. This digital equipment information enables the measurement information to be interpreted correctly, thus a zero reading of motor current can be readily explained if the motor is shown to be switched off.

Another parallel input/output interface provides the **address** and **control** signals necessary for control of the multiplexers, sample/hold and ADC. The microcomputer performs calculations on the input measurement data which enable maximum information on equipment condition to be obtained. Thus, in vibration measurements, Fourier transform calculations can be carried out on displacement velocity and acceleration input data to enable the frequency spectrum of the vibrations to be found. This enables particular vibration frequencies to be associated with certain equipment faults, *e.g.* the failure of a given bearing. The computer converts the condition information into binary coded decimal (bcd) form which is then written into a computer parallel output interface. Each decade is then separately converted into 7-segment code and presented to the plant operator using a 7-segment LED display. The computer also converts each decade of the bcd to ASCII form; the resulting ASCII code is then written into a serial and/or parallel output interface. These can transmit ASCII data in serial and/or parallel form to remote data presentation devices such as a monitor, printer or host computer.

At the beginning of this section, it was stated that there should be ideally a single measured condition or variable associated with a given failure mode in a given item of equipment. In some situations it may be impossible or impracticable to find such a measurement; for example, it may be difficult to measure directly the wear of a given bearing in a centrifugal compressor. However, by measuring a number of other variables associated with the compressor, *e.g.* vibration, temperature and pressure, there should be a given **pattern of variables** associated with the onset of a given failure mode (Section 6.3.2).

Summary

The chapter has analysed the effects of reliability and maintainability on the total cost incurred by the user during the lifetime of the purchased system. This analysis was used to compare the economics of three maintenance strategies; breakdown, preventive and on-condition. The implementation of each of these stagies was then discussed.

References

1. Smith, D. J. (1988) *Reliability and Maintainability in Perspective*, 3rd edn, Macmillan Basingstoke, pp. 25–6.
2. Lees, F. P. (1976) The reliability of instrumentation, *Chemistry and Industry*, March, p. 203.

Self-assessment questions

6.1 Table Q6.1 shows data for two competing liquid level measurement systems. Use the data to decide which system the user should purchase. Assume a breakdown only maintenance strategy and a 10-year total lifetime.

Table Q6.1

Parameter		System (1)	System (2)
Initial cost	(£)	1000	2000
Materials cost per repair	(£)	20	15
Labour cost per hour	(£)	10	10
Process cost per hour	(£)	100	100
Mean down time	(h)	8	12
Annual failure rate	(yr^{-1})	2.0	1.0

6.2 Table Q6.2 gives data for breakdown and on-condition maintenance strategies on a gas compressor. Use the data to decide which strategy should be used.

Table Q6.2 Data for Question 6.2

Parameter	Breakdown maintenance	On-condition maintenance
Initial cost C_I £	200 000	225 000
Materials cost/repair C_R £	20 000	20 000
Repair cost/hour C_L £	50	50
Process cost/hour C_{PB} £	5 000	2 500
MDT hours	60	30
Failure rate λ/yr	0.1	0.1
System lifetime Tyr	10	10
Condition equipment maintenance costs £	0	20 000

6.3 The mean times in hours for maintenance activities associated with the failure and repair of a type of pump are as follows;

    (a) Realization   2
    (b) Access   5
    (c) Diagnosis   2
    (d) Logistics   5
    (e) Repair/replacement 3
    (f) Checkout   2

What is the minimum mean down time corresponding to each of the following maintenance strategies?

(i)   Breakdown
(ii)   Preventive
(iii) On-condition

6.4 For a certain type of high performance tyre, the mean wear-out mileage is 20 000 miles with a standard deviation of 2000 miles. The probability of wear-out before replacement should be 0.1%. Assuming an average mileage of 12 000 miles per year, how frequently should the tyres be replaced?

6.5 In a salt-water environment, cleaning and repainting steelwork extends the life of steelwork by two years. For a steel bridge over a river estuary, 50 000 man-hours of work are required to clean and repaint the entire bridge. How many painters are required for this activity?

6.6 A standby diesel generator has a constant failure rate of 0.15 per year. It is required to have a reliability of at least 0.99; how frequently should the generator be tested?

6.7 In a continuous condition monitoring system for a compressor, the amplitude d mm of displacement vibration of the casing is used to indicate the onset of failure of a given bearing inside the compressor. During the approach of failure d increases with time t in hours according to the equation:

$$d = 2 \times 10^{-9}t^2$$

If potential failure occurs when $d = 0.18$ mm and functional failure when $d = 0.20$ mm, find the potential failure interval.

7

Protective systems for hazardous processes: A case study

7.1 | Introduction

Many plants and processes have an element of hazard associated with them. There is often the risk that system or equipment failure may lead to a local hazard such as an explosion, fire, release of toxic gas or leak of radioactive material. There is a probability that this limited hazard could injure or kill people working in the vicinity, as well as destroying part of the process and causing loss of production. It is established practice that these plants and processes are fitted with automatic protection systems. These systems detect the movement of the process towards a hazardous situation and shut all or part of it down quickly and automatically, thus making it safe without the need for intervention by the process operator.

There are, however, a number of 'super' hazardous plants, which by their very nature present a far greater and wider potential hazard than the 'conventionally' hazardous plants mentioned above. These are often very large, contain large amounts of explosive, poisonous or radioactive material and are operated at high temperatures and pressures. When a fault does occur, a large amount of energy and/or dangerous material can be released over a wide area with large potential loss of life. There have been several such occurrences during the last two decades. At Flixborough, UK in 1974 a vapour cloud explosion at a chemical plant caused 29 deaths. At Ixhuatepic, Mexico in 1984 a liquidified petroleum gas (LPG) explosion caused around 500 deaths. A release of toxic gas (MIC) at a plant at Bhopal, India, also in 1984, caused around 2500 deaths. Finally, a nuclear reactor fire at Chernobyl, USSR in 1986 caused 31 immediate deaths due to radiation sickness together with many hundreds of delayed deaths due to cancer and leukaemia.

Society can only allow such 'super' hazardous plants to be built and operated, if the hazard associated with these processes is greatly reduced to around that of the 'conventionally' hazardous type. In order to achieve this reduction, the 'super' hazardous plant must be protected by a 'super' automatic protective system which is far more comprehensive and reliable than conventional systems. These super systems are referred to as **high integrity protective systems** (HIPS).[1]

This chapter discusses the need for, and design of, a large high integrity protective system for a potentially hazardous chemical reactor and shows how the system meets the target hazard rate.

7.2 | The hazardous process

Figure 7.1(a) is a simplified diagram of the hazardous process.[2] Pure oxygen is reacted with a gaseous hydrocarbon in the presence of a catalyst to produce an oxide of the hydrocarbon. Even with the catalyst, the conversion of hydrocarbon and oxygen to the oxide is low and the unreacted gases are continuously recycled back to the reactors via a recycle gas loop. Upstream of the reactors is the mixing section; here fresh oxygen and hydrocarbon make-up are added to the recycle gas stream. The recycle gas stream also contains unreacted hydrocarbon and oxygen, hydrocarbon oxide, water, carbon dioxide and inert gases. In the reactor two competing reactions take place simultaneously. In the first wanted reaction, there is limited oxidation of the hydrocarbon to give the hydrocarbon oxide and a small amount of heat is produced. In the second unwanted reaction, there is complete oxidation of the hydrocarbon to carbon dioxide and water with the production of a large amount of heat. The second reaction uses far more oxygen than the first, so that the main hazard is excessive oxygen concentration. Thus, if the oxygen concentration at any point in the process is below a critical percentage called the **flammable limit**, then the first reaction predominates and the plant is safe. If the oxygen concentration is above this flammable limit, then the second reaction predominates with large, and potentially highly explosive generation of heat, and the plant is unsafe (Figure 7.1(b)). In this situation the action required to make the plant safe again is clear; the oxygen make-up supply should be shut off automatically and quickly to bring the oxygen concentration back below the flammable limit. The gas stream leaving the reactors is then cooled to condense the hydrocarbon oxide and water: this liquid mixture is then removed from the gas stream in the product separation stage and passes to the product purification stage where the hydrocarbon oxide is extracted. The remaining gas stream then passes to the recycle gas compressor; here it is compressed prior to passing to the by-product removal stage, where carbon dioxide and steam are removed. The gas then passes to the mixing section and the description of the process is therefore complete.

7.3 | Target hazard rate for the protective system

Having established that the process is of the 'super' hazardous type, it will need to be protected by a high integrity protective system. The following sections discuss the design

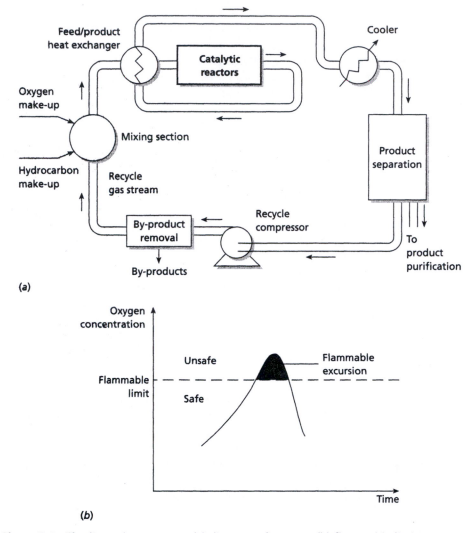

Figure 7.1 The hazardous process: (a) diagram of process; (b) flammable limit

of a suitable HIPS. The first stage in the design of any system or item of equipment is to establish a design specification; the design specification for the HIPS is a numerical value for the **target hazard rate**. To calculate this value the **background risk** must first be established; this is the mean **fatal accident rate** (FAR) on all of the chemical plants operated by the company concerned. The FAR was found to be 3.5 deaths in 10^8 exposed hours. To put this figure into perspective, Table 7.1 gives some corresponding FAR figures for different industries and occupations in the United Kingdom and some non-industrial activities. Assuming continuous operation of the plants for 8700 hours in any year, then the mean probability of a fatality on any one of the company's plants is

Table 7.1 Fatal accident rates

	Fatal accident rate (FAR) (deaths/10^8 exposed hours)
Fatal accident rates in different industries and jobs in the UK	
Clothing and footwear industry	0.15
Vehicle industry	1.3
Chemical industry	5
British industry	4
Steel industry	8
Agricultural work	10
Fishing	35
Construction work	67
Air crew	250
Professional boxers	7 000
Jockeys (flat racing)	50 000
Fatal accident rates for the chemical industry in different countries	
France	8.5
West Germany	5
United Kingdom (before Flixborough)	4
(including Flixborough)	5
Fatal accident rates for some non-industrial activities	
Staying at home	3
Travelling:	
by bus	3
by train	5
by car	57
by bicycle	96
by air	240
by moped	260
by motor scooter	310
by motor cycle	660
Canoeing	1 000
Rock climbing	4 000

$3.5 \times 10^{-8} \times 8700 = 3 \times 10^{-4}$ per year. This is the **background risk** for any employee on any of the company's plants. If the company expects the employee to work on a 'super' hazardous plant, then the **additional risk** should be no more than $\frac{1}{10}$th of the background risk, *i.e.* 3×10^{-5} per year. Since the HIPS system is required to keep the additional risk to 3×10^{-5} per year, this is the numerical value for the **target hazard rate**.

7.4 | Choice of trip initiating parameters

The process of quickly and automatically shutting off the oxygen make-up supply to the plant is called a 'trip'. Trip initiating parameters are measured variables associated with

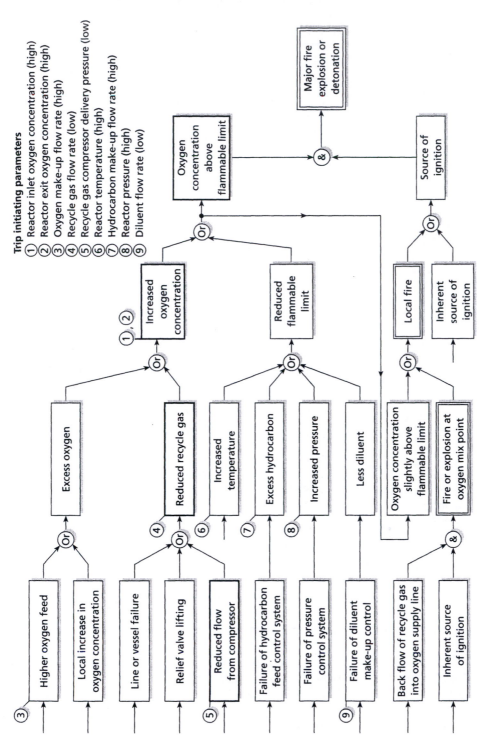

Trip initiating parameters

① Reactor inlet oxygen concentration (high)
② Reactor exit oxygen concentration (high)
③ Oxygen make-up flow rate (high)
④ Recycle gas flow rate (low)
⑤ Recycle gas compressor delivery pressure (low)
⑥ Reactor temperature (high)
⑦ Hydrocarbon make-up flow rate (high)
⑧ Reactor pressure (high)
⑨ Diluent flow rate (low)

Figure 7.2 Section of fault tree (after Stewart[1])

the plant, which indicate that the plant may be tending towards a potentially hazardous condition and can be used to initiate a trip. In a HIPS it is essential that all process conditions and equipment faults that could either:

(a)   directly or
(b)   in conjunction with other faults

cause the major hazard, are monitored by trip initiating parameters. Some parameters are obvious; high oxygen concentration has already been mentioned in Section 7.2; high reactor temperature and high reactor pressure are others. Other parameters are far less obvious and are required either to protect against a highly unlikely combination of faults, or because the measurement of other more obvious parameters (*e.g.* oxygen concentration) is too slow, or because the obvious parameters initiate trips whose action could result in a hazardous condition (*e.g.* recycle gas compressor trips).

Fault tree analysis (Section 5.5.2) provides a systematic way of identifying all potentially hazardous conditions and equipment faults. Figure 7.2 shows a small section of the fault tree for the hydrocarbon–oxygen process;[1] we see that the diagram 'fans out' into a whole series of hazardous conditions and faults. Thus oxygen concentration above the flammable limit is caused by increased oxygen concentration OR reduced flammable limit. Reduced flammable limit is in turn caused by increased temperature OR excess hydrocarbon OR increased pressure OR less diluent. There must be at least one trip initiating parameter associated with each line of logic. Figure 7.2 shows nine trip initiating parameters associated with this section of the fault tree; there are around 50 associated with the full fault tree.

7.5 │ Design of the protective system

7.5.1 Basic trip systems

Before considering the detailed design of the HIPS, we first consider the design and reliability of a basic trip system that would be used to protect a 'conventionally' hazardous plant. A typical simple trip system is shown in Figure 7.3; it consists of a switch, a three-way solenoid valve and a pneumatically operated trip valve. The position of the switch is set by comparing the value of the trip initiating parameter X (*e.g.* pressure, temperature and flow rate) with the value X_T, at which a plant trip is required (trip setting). Consider the example of a high pressure trip with a trip setting of 25 bar; if the pressure is less than 25 bar, the plant is in a normal, safe condition and the switch is closed. If the pressure is greater than 25 bar, then the plant is in an unsafe condition and the switch is open. The switch is in a series electrical circuit with the solenoid and a 24 V DC power supply. Thus, under normal conditions the switch is closed and the solenoid energized; under trip conditions the switch is open and the solenoid is de-energized. The solenoid valve has three ports 'in', 'out' and 'vent', and is placed in the air supply line to a pneumatically operated trip valve. Under normal conditions when the solenoid is energized,

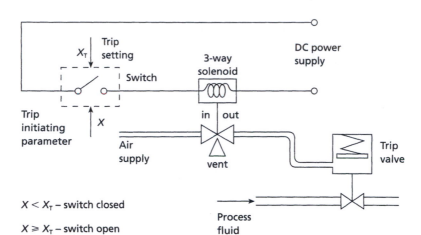

Figure 7.3   Basic trip system

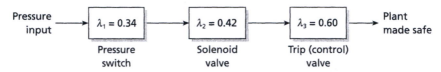

Figure 7.4   Block diagram of trip system

the 'vent' port is closed, air flows from the 'in' port to the 'out' port and the trip valve has its normal air supply. Under trip conditions when the solenoid is de-energized, the 'in' port is closed and air flows from the 'out' port to the 'vent' port. This means that air is vented from the bonnet of the trip valve to the atmosphere causing the air pressure in the bonnet to fall. If the trip valve must close to make the plant safe (*e.g.* shutting off the supply of oxygen) then a 'close air failure' valve is installed which closes when the air pressure falls. If the trip valve must open to make the plant safe (*e.g.* increasing the supply of cooling water) then an 'open air failure' valve is installed which opens when the air pressure falls.

Figure 7.4 is a block diagram representation of the trip system; it consists of three elements in series with annual failure rates λ_1, λ_2 and λ_3. From Equation 3.5 the corresponding system annual failure rate λ is given by:

$$\lambda = \lambda_1 + \lambda_2 + \lambda_3 = 1.36 \tag{7.1}$$

All three element failure rates are assumed to be constant with time (Section 2.5) and the numerical values are typical for instrumentation in a chemical plant environment. The probability of a system failure is therefore given by the exponential relation (Equation 2.16):

$$F(t) = 1 - e^{-\lambda t} = 1 - e^{-1.36t} \tag{7.2}$$

Thus, when $t = 0.5$ year, $F = 0.493$; this is far too high to be acceptable in a trip system and must be reduced.

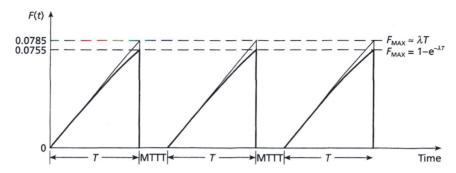

Figure 7.5 Time variation of F with proof testing

One way of reducing F is **proof testing** or **trip testing**. This is the regular, periodic testing of the complete trip system in order to reveal any hidden faults and has already been discussed in Section 6.4.3. The trip or proof testing interval T is the interval between successive proof tests on the system. Mean time to test (MTTT) is the mean time taken to carry out the proof test. Just after the system has been tested and proved to be working correctly the unreliability is $F = 0$. Thereafter, the unreliability increases exponentially to the maximum value $F_{MAX} = 1 - e^{-\lambda T}$ just before the next trip test. The time variation of F has therefore the sawtooth waveform shown in Figure 7.5. A typical trip testing interval would be $T = 3$ weeks $= 3/52$ year $= 0.0577$ year. Therefore, $\lambda T = 0.0785$ and $F_{MAX} = 0.0755$. The approximation $1 - e^{-\lambda T} \approx \lambda T = 0.0785$ is therefore valid in trip systems, where λT is small compared with 1. The value of F_{MAX} may still be too large for 'super' hazardous plants.

Before discussing further strategies for reducing the probability of system failure, we must first recognize that there are two distinct types of failure in protective systems. A **fail-danger failure** is any system or component failure which prevents, or tends to prevent, the plant being tripped when a potentially hazardous fault or condition occurs. An example of fail-danger failure would occur if a pressure switch failed to open when the pressure exceeded the trip pressure. A **fail-safe failure** is any system or component failure which produces a plant trip when a plant trip is not required. In the above example, a fail-safe failure would occur if the pressure switch opened when the pressure was below the trip pressure. A fail-danger failure is a very serious occurrence; it prevents the trip system from protecting the plant when a hazard may occur. Fail-safe failures are less serious but unnecessary trips cause lost production and loss of management and operator confidence in the trip system.

In general, fail-danger and fail-safe failures will have different failure rates λ_D and λ_S respectively, and therefore different failure probabilities:

$$F_D = 1 - e^{-\lambda_D t}, \quad F_S = 1 - e^{-\lambda_S t} \tag{7.3}$$

However, detailed information on the failure rates associated with all possible modes of failure of trip equipment is not always available. We may have to assume that λ_D and λ_S are both equal to the value quoted for average failure rate. One exception is the data

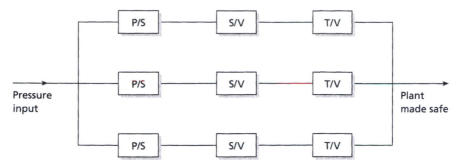

Figure 7.6   Three trip systems in parallel

given by Taylor for the failure rates of the trip system instrumentation on a chemical plant.[3]

Fractional dead time FDT is closely related to **fail-danger probability**. This is the mean proportion of the proof testing interval T that the trip system is in a fail-danger state, *i.e.* the mean proportion of T that it is incapable of protecting the plant, *i.e.* 'dead'. Thus we have:

$$\text{FDT} = \frac{1}{T} \int_0^T F_D(t)\,dt \tag{7.4}$$

Thus fractional dead time is a similar concept to unavailability (see Section 3.8).

Another possible method for reducing system failure probability is **redundancy** (Section 3.3). One example is the use of several identical trip systems in parallel all using the **same trip initiating parameter**, *e.g.* three identical temperature trip systems in parallel. Consider an overall trip system consisting of three of the above pressure trip systems or channels in parallel, *i.e.* the arrangement of Figure 7.6.

If any single channel is in a fail-danger condition then the other two channels are still able to protect the plant. Thus **the overall system fail-danger probability P_D is less** than the single channel fail-danger probability F_D; using Equation 3.9 we have:

Overall system fail-danger probability $P_D = F_D^3$ $\qquad$ (7.5)

Assuming a three-week trip testing interval and $\lambda_D = 1.36$ faults/channel/year, we showed earlier that the maximum value of F_D for a single channel was 0.0755. Thus:

$$P_D = (0.0755)^3 = 4.30 \times 10^{-4}$$

i.e. a considerable reduction in fail-danger probability. However, the **overall system fail-safe probability P_S is greater** than the single channel fail-safe probability F_S. If one channel is in a fail-safe condition the plant is tripped anyway, even though the other two may be working correctly. Since there are three channels and a fail-safe condition in any one will trip the plant, then the overall system fail-safe probability is three times that for the single channel.

Overall system fail-safe probability

$$P_S = 3F_S \tag{7.6}$$

Assuming maximum F_S for a single channel $= F_D = 0.0755$ then $P_S = 0.227$. This is far too high to be acceptable to the plant management: there would be far too many unnecessary trips resulting in lost production and profit. We require therefore redundant trip systems for which both fail-danger and fail-safe probability are acceptable. This can be achieved using majority voting systems.

7.5.2 Majority voting systems

In a majority voting system there are n trip channels in parallel and the plant is tripped if m ($m \leqslant n$) channels indicate that the plant should be tripped. Such a system is referred to as 'm out of n' or m oo n. Majority voting systems have already been discussed in Section 3.5 and we saw how the binomial distribution can be used to calculate overall failure probabilities. Supposing we wish to calculate overall fail-danger and fail-safe probabilities for a system with 'two out of three' voting, $i.e.$ 2 oo 3 where $m = 2$, $n = 3$. These probabilities can be calculated from the binomial expansion of $(R + F)^3$, where R and F are the single channel reliability and unreliability respectively. Thus we have:

$$(R + F)^3 = F^3 + 3RF^2 + 3R^2F + R^3 \tag{7.7}$$

where F^3 represents the probability of all three channels failing, $3RF^2$ the probability of two channels failing, $3R^2F$ the probability of one channel failing and R^3 the probability of no channels failing.

Looking at fail-danger first, if either two or three channels fail dangerously then there are correspondingly only one or zero channels left working. This is insufficient to trip the plant with 2 oo 3 voting and an overall fail-danger failure has occurred. If F_D is the single channel fail danger probability, then the overall system fail-danger probability is:

$$P_D = F_D^3 + 3RF_D^2 = F_D^2(3R + F_D)$$

In a protective system R will be close to 1 and F_D will be a lot smaller than 1 giving:

$$P_D \approx 3F_D^2 \tag{7.8}$$

Looking at fail-safe, a fail-safe failure of no channels or only one channel will not cause a plant trip with 2 oo 3 voting. A fail-safe failure of two channels will cause an unnecessary plant trip; the failure of a third channel is irrelevant because the plant is tripped by only two channels. The overall system fail-safe probability is therefore:

$$P_S = 3RF_S^2 \approx 3F_S^2 \tag{7.9}$$

where F_S is the single channel fail-safe probability.

We note that for 2 oo 3 voting the algebraic expressions for P_D and P_S are identical. Assuming that both F_D and F_S have maximum values of 0.0755 then both P_D and P_S are equal to 0.017. We can draw two further conclusions about 2 oo 3 voting:

(a) The fail-safe probability for 2 oo 3 voting is lower than for either the single channel (1 oo 1: 0.0755) or three channels in parallel (1 oo 3: 0.227).

(b) The fail-danger probability for 2 oo 3 voting is lower than for 1 oo 1 (0.0755) but much greater than for 1 oo 3 (4.30×10^{-4}).

General expressions for overall fail-danger probability, P_D, fail-safe probability, P_S, and fractional dead time, FDT, for m oo n systems are derived in Appendix B. These are:

Overall fail-danger probability

$$P_D = {}^nC_r F_D^r \qquad (7.10)$$

where $r = n - m + 1$

$${}^nC_r = \frac{n!}{r!(n-r)!}$$

F_D = single channel fail-danger probability

$F_D^{MAX} = 1 - e^{-\lambda_D T} \approx \lambda_D T \quad \text{(if } \lambda_D T \ll 1)$

Overall fail-safe probability

$$P_S = {}^nC_m F_S^m \qquad (7.11)$$

where ${}^nC_m = \dfrac{n!}{m!(n-m)!}$

F_S = single channel fail-safe probability

$F_S^{MAX} = 1 - e^{-\lambda_S T} \approx \lambda_S T \text{ (if } \lambda_S T \ll 1)$

Fractional dead time FDT

$$\text{FDT} = \frac{1}{r+1} {}^nC_r F_D^r \qquad (7.12)$$

Table 7.2 gives detailed algebraic expressions for the above three quantities.

Majority voting can be implemented with combinational logic. Figure 7.7(a) shows that a 2 oo 3 voting element will have three input logic signals A, B, C and a single logic output signal Z. Each input signal is derived from a trip initiating instrument, which compares the value of the trip initiating parameter with the trip setting (Section 7.5.1). In a normal situation the instrument gives a logic '0' output, in a trip situation a logic '1' output. The element output signal Z is '0' in a normal situation and '1' if a trip is required; this signal is then used to initiate a shut down. Figure 7.7(a) shows the truth table for 2 oo 3 voting; the corresponding Boolean expression is:

Table 7.2 Detailed algebraic expressions for P_D, P_S and FDT

m oo n	r = n − m + 1	Fail-danger probability $P_D = {}^nC_rF_D^r$	Fail-safe probability $P_S = nC_mF_S^m$	$FDT = \dfrac{1}{r+1}C_rF_D^r$
1 oo 1	1	F_D	F_S	$\frac{1}{2}F_D$
1 oo 2	2	F_D^2	$2F_S$	$\frac{1}{3}F_D^2$
2 oo 2	1	$2F_D$	F_S^2	F_D
1 oo 3	3	F_D^3	$3F_S$	$\frac{1}{4}F_D^3$
2 oo 3	2	$3F_D^2$	$3F_S^2$	F_D^2
3 oo 3	1	$3F_D$	F_S^3	$\frac{3}{2}F_D$
1 oo 4	4	F_D^4	$4F_S$	$\frac{1}{5}F_D$
2 oo 4	3	$4F_D^3$	$6F_S^2$	F_D^3
3 oo 4	2	$6F_D^2$	$4F_S^3$	$2F_D^2$
4 oo 4	1	$4F_D$	F_S^4	$2F_D$
1 oo 5	5	F_D^5	$5F_S$	$\frac{1}{6}F_D^5$
2 oo 5	4	$5F_D^4$	$10F_S^2$	F_D^4
3 oo 5	3	$10F_D^3$	$10F_S^3$	$\frac{5}{2}F_D^3$
4 oo 5	2	$10F_D^2$	$5F_S^4$	$\frac{10}{3}F_D^2$

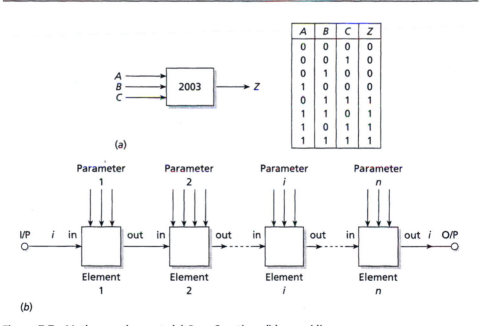

Figure 7.7 Voting equipment: (a) 2 oo 3 voting; (b) guard line

$$Z = (A + C)(B + A)(C + B) \tag{7.13}$$

So far we have seen the need for **redundancy** in protective systems; this is the use of several identical trip channels all with the same trip initiating parameter. Also in Section 7.4 we saw the need for many trip initiating parameters; this is **diversity**. A high integrity protective system must have both redundancy and diversity, *e.g.* three temperature trip channels, three pressure trip channels, three oxygen concentration trip channels, *etc.*, in order to meet the specified target hazard rate. The use of **guard lines** enables both features to be obtained. A guard line consists of a number n of interconnected voting elements, one for each trip initiating parameter. The basic requirement is that if any individual voting element is in a trip condition then the entire guard line is tripped. This means that the voting elements must be connected together in *series* (Section 3.2). This series interconnection is obtained by providing each element with electrical input and output ports (Figure 7.7(b)). In normal operation a current i is transmitted along the entire guard line from input to output; if, however, one element trips then the current flow ceases and the whole guard line trips.

7.5.3 Detail of the protective system

Figure 7.8 shows an overview of the plant and high integrity protective system.[1,2] There are n different trip initiating plant parameters. The **high integrity trip initiators** (HITIs)

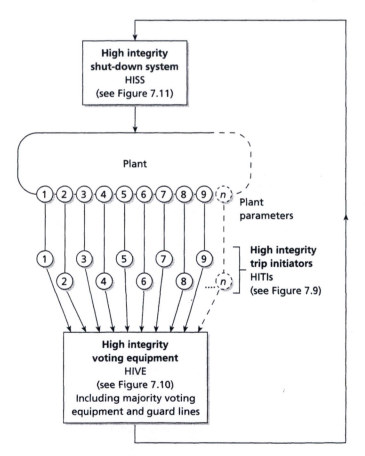

Figure 7.8   Overview of plant and high integrity protective system (after Stewart[1])

compare the measured value of each parameter with the trip setting and give logic output signals. These are passed to the **high integrity voting equipment** (HIVE) which uses majority voting to decide whether a plant trip is required. The 'trip' or 'no trip' signal is then passed to the **high integrity shutdown system** (HISS); if a trip is required this shuts off the oxygen make-up supply to make the plant safe.

Figure 7.9 shows a section of the HITIs; to ensure adequate redundancy there are three HITIs for each parameter.[1,2] Each initiator consists of a transducer, which produces

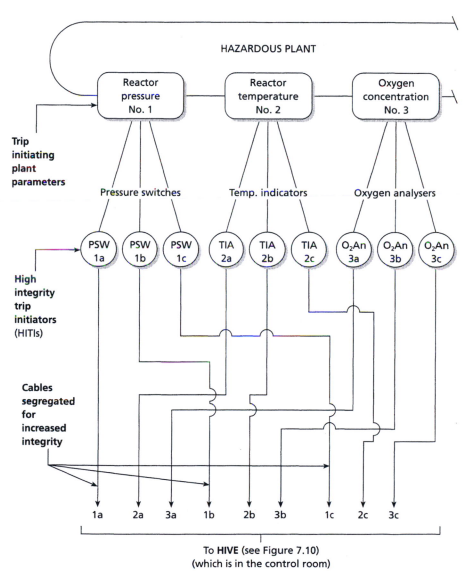

Figure 7.9   High integrity trip initiators (HITIs) (after Stewart[1])

a signal corresponding to the current value of the parameter and a comparator which compares this signal value with the trip setting and gives a corresponding 0 or 1 logic output signal. Each HITI must fulfill two main requirements:

(a) have a much lower than average failure rate, typically λ should be around 0.1 per year;

(b) have a fast speed of response.

To ensure that the above redundancy is not compromised, the following requirements should also be satisfied.

(c) Each initiating instrument must have a separate power supply.

(d) Signal cables from each of the three HITIs must take different routes to the control room. For example, the cable from pressure switch PSW 1a takes the 'a' route, the cable from PSW 1b the 'b' route, the cable from PSW 1c the 'c' route.

(e) Each of the three HITIs should be proof tested at different times. If a three-week trip testing interval is used, then PSW 1a will be tested during week 1, PSW 1b during week 2, PSW 1c during week 3, returning to PSW 1a during week 4.

Figure 7.10 shows the layout of the HIVE.[1,2] Two out of three voting was chosen for the following reasons:

(a) The fail-safe failure of one HITI does not trip the plant, giving fewer unnecessary trips. We saw in Section 7.5.2 that the fail-safe probability for 2 oo 3 voting is lower than for either a single channel (1 oo 1) or three channels in parallel (1 oo 3).

(b) The fail-danger failure of one HITI does not prevent the plant being tripped when it should. From Section 7.5.2 the fail-danger probability of 2 oo 3 is lower than for 1 oo 1 but greater than for 1 oo 3.

(c) It allows testing of the HITIs without having either to shut down the plant or disarm them. One HITI is tested at a time and the plant is still protected by the other two HITIs.

There is one 2 oo 3 element for each parameter; if there are n parameters, n elements are connected in series to form a guard line. There are three identical guard lines; the output of each guard line goes to another 2 oo 3 voting element, and the output signal of this element passes to the shut-down system. Two out of three voting of guard lines again gives lower fail-danger and fail-safe failure probabilities than a single guard line. The shut-down system is duplicated to allow one half of the shut-down system to be tested, while the other half protects the plant; it is also necessary to duplicate the HIVE. There is an 'A' side with guard lines A1, A2, A3 and an identical 'B' side with guard lines B1, B2, B3. Each of the three HITI input signal cables for a given parameter, *e.g.* 1a, 1b, 1c, is 'fanned out' to six output signal lines. Each of these output lines passes to the voting element associated with that parameter in each of the six guard lines. This means that the three input signals to each voting element have come via the three different routes a, b, c. The HIVE provides two output logic signals to the shut-down system, the A signal from the final voting element for the A guard lines and the B signal from the final voting element for the B guard lines.

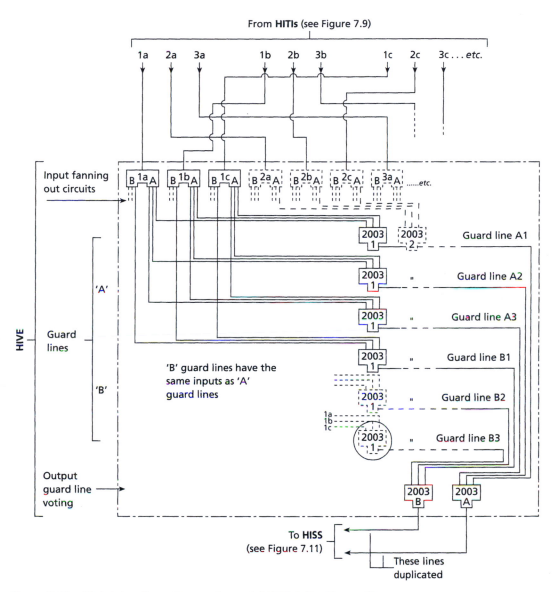

Figure 7.10 High integrity voting equipment (HIVE) (after Stewart[1])

Figure 7.11 shows the high integrity shut-down system, HISS;[1,2] the HISS is designed to cut off the supply of oxygen to the plant completely, quickly and safely in response to a trip signal from the HIVE. The oxygen enters the plant through a large bore pipe; it is therefore imperative that the overall fail-danger probability P_D of the trip valve system be as low as possible. For a single valve system, P_D is equal to the single channel fail-danger probability F_D; this is far too high. A system of two trip valves in series with their intervalve space vented was then considered. The valves are in series hydraulically,

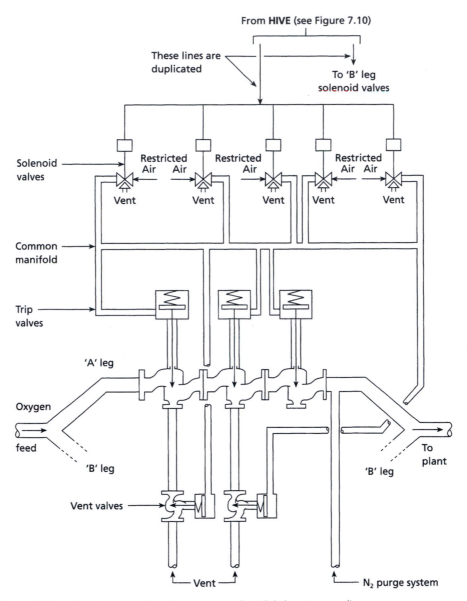

Figure 7.11   High integrity shutdown system (HISS) (after Stewart[1])

i.e. the same flow passes through both. From a reliability point of view, however, the valves are in parallel, *i.e.* a 1 oo 2 system with $P_D = F_D^2$. P_D was again found to be too high so that a three-valve system, *i.e.* 1 oo 3 with $P_D = F_D^3$, was found to be necessary. The other requirement of the shut-down system is that it must be possible to proof test it online, without shutting off the oxygen feed to the plant. The shut-down system is

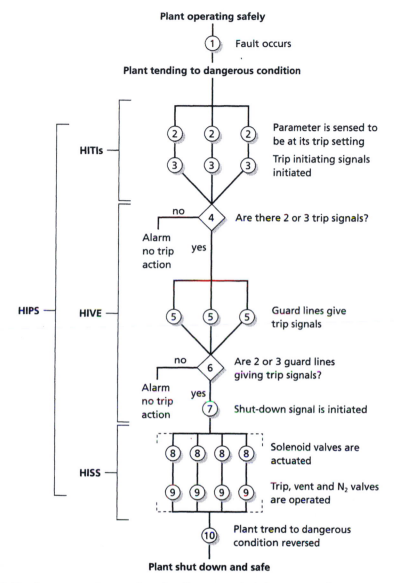

Figure 7.12 Sequence of operations leading to a trip (after Stewart[1])

therefore duplicated, *i.e.* an A leg with three valves and an identical B leg with three valves. Thus the A side of the voting and shut-down systems can be tested while the plant is both run and protected by the B side and vice versa.

The operation of the complete protective system is shown in Figure 7.12.[1,2] If, for any reason, the process is tending towards a dangerous condition, then this is detected by (at least two of) the trip initiating parameters. When a parameter reaches its trip setting, its three HITIs each send a trip initiating signal to the HIVE; if only one HITI sends such a signal there is only an alarm, but if two out of three send out trip initiating signals the HIVE's guard lines give trip signals. Provided at least two of the three guard lines in the HIVE are giving trip signals, then a shut-down signal is sent to the HISS. The solenoid valves are then activated, the trip valves shut, vent valves opened, the oxygen flow is stopped and the plant is made safe.

7.6 | Hazard analysis of process and protective system

Once the hazardous process and its protective system have been designed, a **hazard analysis** should be carried out to ensure that the **actual hazard rate** is within the **target hazard rate** value of 3×10^{-5} specified in Section 7.3.

Hazard rate is simply the number of hazards that occur in a given time interval, *e.g.* one year. To illustrate the method of calculating hazard rate, we consider the example of braking system failure, which prevents a car stopping suddenly in an emergency situation. The average number of times in a year that the car is required to stop suddenly is called the **demand rate** (DR), let us assume DR = 2.0 per year. The mean unreliability of the braking system can be quantified using **fractional dead time** (FDT). This is the mean proportion of a given time interval (*e.g.* one year) that the braking system is 'dead', *i.e.* incapable of stopping the car suddenly on demand (Section 7.5). Suppose that the FDT for the braking system is 10^{-3}. A hazard results if a demand coincides with this dead period, *i.e.* if there is a demand **and** the system is 'dead'. Hazard rate is therefore the **product** of demand rate and fractional dead time. The resulting annual hazard rate in this example is 2×10^{-3} and in general we have:

hazard rate = demand rate × fractional dead time
$$= (DR) \times (FDT) \tag{7.14}$$

Equation 7.14 is generally applicable to any hazardous situation where FDT is small. It can therefore be used to calculate the actual hazard rate for the process and protective system.

The first stage is to identify all process conditions and equipment faults that can lead directly to the major hazard. This has already been done using fault-tree analysis (Section 7.4 and Figure 7.2). The total achieved hazard rate H is then the sum of the individual hazard rates due to all possible fault conditions, *i.e.*

$$H = \sum_{i=1}^{i=N} H_i \tag{7.15}$$

where H_i = hazard rate associated with the ith fault condition
N = total number of all possible fault conditions.

Using Equation 7.14 we have:

$$H_i = (DR)_i \times (FDT)_i \tag{7.16}$$

Here $(DR)_i$ is the demand rate the ith fault condition places on the ith system, which protects the plant from that fault condition. $(FDT)_i$ is the corresponding fractional dead time for the ith protective system. Equation 7.16 emphasizes the importance of having a protective system for each fault condition; if there is no such system $(FDT)_i = 1$ and $H_i = (DR)_i$.

In general $(DR)_i$ is not equal to the number of occasions/year the fault occurs. We must also calculate the probability that the fault does eventually result in a major explosion or rupture in the reactor. This probability may not necessarily be 1.0; for example, if the fault occurs some distance away from the reactor. We must also calculate the probability that operator intervention fails. If the fault results in a slow movement towards a hazardous condition, then the operator may be able to shut off the supply of oxygen manually and make the plant safe. The above probability is again less than one and $(DR)_i$ is correspondingly reduced. We therefore have:

$$
\begin{aligned}
(DR)_i = {} & \text{Number of occasions per year } i\text{th fault occurs} \\
& \times \text{Probability that it leads to rupture} \\
& \times \text{Probability that manual intervention fails}
\end{aligned}
\tag{7.17}
$$

The ith protective system can be considered to be made of three elements, initiation, voting and shut-down, arranged in series. In Section 3.2 we saw that provided the **individual element failure probabilities F_i are very small**, then the system failure probability F_{SYST} is approximately equal to the sum of the F_i, *i.e.*

$$
F_{SYST} \approx \sum_{i=1}^{i=s} F_i
\tag{3.7}
$$

Since FDT is closely related to fail-danger probability F_D and since FDT $\ll 1$ for a protective system, the fractional dead time for the ith system is approximately the sum of the element FDTs, *i.e.*

$$
(FDT)_i = (FDT)_i^{HITI} + (FDT)_i^{HIVE} + (FDT)_i^{HISS}
\tag{7.18}
$$

The individual element FDT values in the general case of m oo n voting are given by Equation 7.12, *i.e.*:

$$
FDT = \frac{1}{r+1}{}^n C_r (\lambda_D T)^r
\tag{7.12}
$$

where $r = n - m + 1$, λ_D = fail-danger failure rate and T = trip testing interval (see Appendix B for derivation and Table 7.2 for algebraic expressions). Thus for the HITIs with 2 oo 3 voting we have FDT = $(\lambda_D T)^2$, for a single voting element, *i.e.* 1 oo 1, we have FDT = $\frac{1}{2}(\lambda_D T)$ and for three shut-down elements in parallel, *i.e.* 1 oo 3, we have FDT = $\frac{1}{4}(\lambda_D T)^3$.

Table 7.3 shows part of the calculation of achieved hazard rate for the hydrocarbon oxide reactor and the associated protective system.[1,2] Five potential fault conditions are shown and the corresponding demand rates (columns (a) to (d)) and fractional dead times

Table 7.3 Section of hazard rate calculation (after Stewart[1])

Description	Fault condition			d ∴ Demand rate ($=a \times b \times c$)	Relevant trip initiator No.	Fractional dead times				Hazard rate × 10^5 ($= d \times h$)
	a Occasions per year	b Probability that it leads to rupture	c Probability that operator's intervention fails			e HITI	f HIVE	g HISS	h ∴ Overall ($= e + f + g$)	
1. Feed filters blocked	0.001	0.2	0.1	0.000 02	10 and 12	10^{-4}	10^{-5}	10^{-5}	1.2×10^{-4}	0.000 24
2. Oxygen supply failure	2.0	0.2	0.1	0.04	10 and 38	10^{-4}	10^{-5}	10^{-5}	1.2×10^{-4}	0.48
3. PCV fails open	0.25	0.1	1.0	0.025	11	10^{-3}	8.3×10^{-5}	10^{-5}	1.09×10^{-3}	2.72
4. Compressor anti-surge bypass fails to open	0.2	1.0	0.1	0.02	18 and 36	10^{-4}	10^{-5}	10^{-5}	1.2×10^{-4}	0.24
5. Gross carry-over from absorber	0.1	1.0	0.1	0.01	18, 24	10^{-4}	10^{-5}	10^{-5}	1.2×10^{-4}	0.12
⋮	⋮	⋮	⋮	⋮	⋮	⋮	⋮	⋮	⋮	⋮

(columns (e) to (h)) calculated. The final column gives the individual hazard rates H_i (multiplied by 10^5). The total actual hazard rate (*i.e.* sum of *all* individual hazard rates is 4.79×10^{-5}. This is clearly greater than the **target hazard rate** of 3×10^{-5} and must be reduced. The largest contributor is the hazard rate of 2.72×10^{-5} associated with fault condition 3, the failure open of the pressure control valve in the oxygen supply line to the plant: this must be reduced first. When the valve fails open, the flow rate of oxygen increases rapidly causing the oxygen concentration to increase above the flammable limit. Trip initiator No. 11, high oxygen flow rate, is used to protect the plant from this condition. However, a considerable flammable excursion could occur before this trip is initiated. An extra trip initiator, based on the high rate of rise of oxygen flow rate, was therefore added to protect against fault condition 3. This reduced the individual hazard rate to 0.8×10^{-5} and the total hazard rate to 2.87×10^{-5} which is less than the target hazard rate.

Summary

This chapter has shown that in order to protect a potentially hazardous plant a high integrity protective system, which meets a target hazard rate, is required. The chapter then discussed the detailed design of such a system and demonstrated how the achieved hazard rate meets the above target.

References

1. Stewart, R. M. (1971) High integrity protective systems, *Institution of Chemical Engineers Symposium Series*, **34**, pp. 99–104.
2. Stewart, R. M. (1974) The design and operation of high integrity protective systems, *Process Safety – Theory and Practice*, July, Teesside Polytechnic.
3. Taylor, A. (1980) Comparison of predicted and actual reliabilities on a chemical plant, *I.Chem.E. Symposium Series*, **66**, pp. 105–16.

Worked examples

Example 7.1: Design of a protective system

A protective system, based on temperature measurement, is to have a maximum fail-danger probability not exceeding 8×10^{-3} and a maximum fail-safe probability not exceeding 5×10^{-2}. The system is tested and proved to be working correctly at three-week intervals. Use the components listed below to design a protective system which meets the above specification. Annual fail-safe and fail-danger failure rates for each component are as shown. Full justification and explanation should be given.

Thermocouple $\lambda_S = \lambda_D = 0.5$
Thermocouple input trip amplifier/comparator
   (switch operation or logic output) $\lambda_S = \lambda_D = 0.1$
m out of n voting element (logic input and output signals) $\lambda_S = \lambda_D = 0.05$
Logic operated switch (open or closed depending
   on logic level) $\lambda_S = \lambda_D = 0.1$
Three-way solenoid valve (in air supply line) $\lambda_S = \lambda_D = 0.1$
Trip valve (pneumatic) $\lambda_S = \lambda_D = 0.1$

Solutions

Basic single channel system

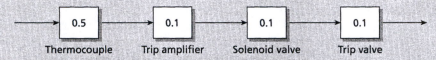

Figure 7.13   Basic single channel system of Example 7.1

System failure rate $\lambda_S = 0.5 + 0.1 + 0.1 + 0.1 = 0.8/\text{year}$

$$F_S = F_D = 1 - e{-}\lambda_S t,$$

these are maximum when $t = T = 3/52$ year

$$F_S^{MAX} = F_D^{MAX} = 1 - e^{-0.8 \times 3/52} = \mathbf{4.51 \times 10^{-2}}$$

Fail-safe OK (just), fail-danger too high.

Several single channels in parallel

2 in parallel

$$P_D = F_D^2 = (4.51)^2 \times 10^{-4} \approx \mathbf{2 \times 10^{-3}} \quad - \text{OK}$$

$$P_S = 2F_S = 2 \times 4.51 \times 10^{-2} \approx \mathbf{9 \times 10^{-2}} \quad - \text{Too high}$$

3 in parallel

$$P_D = F_D^3 = (4.51)^3 \times 10^{-6} \approx \mathbf{9.2 \times 10^{-5}} \quad - \text{OK}$$

$$P_S = 3F_S = 3 \times 4.51 \times 10^{-2} \approx \mathbf{14 \times 10^{-2}} \quad - \text{Even worse}$$

Need to use majority voting system:

Most likely structure:

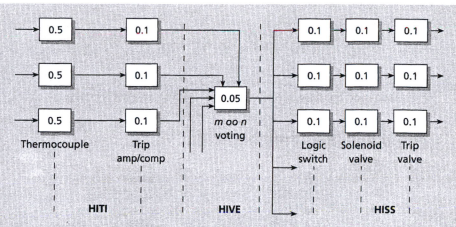

Figure 7.14 Several single channels in parallel

HIVE

Low failure rate – need not duplicate. **Single element $\lambda = 0.05$**

$\quad$ Maximum $P_S = P_D = F_S = F_D = 1 - e^{-0.05 \times 3/52} = \mathbf{2.8 \times 10^{-3}}$

Need to reduce other contributions to this level.

HITI

Single channel $\lambda = 0.6$

$\quad$ Maximum $F_S = F_D = 1 - e^{-0.6 \times 3/52} = \mathbf{3.4 \times 10^{-2}}$

Need to reduce.

2 oo 3 voting

$P_D = 3F_D^2 = 3 \times (3.4 \times 10^{-2})^2 \approx \mathbf{3.4 \times 10^{-3}}$
$P_S = 3F_S^2 \approx \mathbf{3.4 \times 10^{-3}}$

P_S, P_D are now comparable with HIVE.

HISS

Single channel $\lambda = 0.3$

$\quad$ Maximum $F_S = F_D = 1 - e^{-0.3 \times 3/52} \approx \mathbf{1.7 \times 10^{-2}}$

Need to reduce F_D without increasing F_S too much.

Two channels in parallel

$$P_D = F_D^2 = (1.7 \times 10^{-2})^2 \approx \textbf{0.3} \times \textbf{10}^{-3} \qquad - \text{OK}$$

$$P_S = 2F_S = 2 \times 1.7 \times 10^{-2} = \textbf{3.44} \times \textbf{10}^{-2}$$

P_S is increased but may still meet overall specification for system.

Total system failure probabilities

Fail-danger

$$
\begin{aligned}
\text{Probability} &= (P_D)_{\text{HITI}} + (P_D)_{\text{HIVE}} + (P_D)_{\text{HISS}} \\
&= 3.4 \times 10^{-3} + 2.8 \times 10^{-3} + 0.3 \times 10^{-3} \\
&\quad\ \ (2\text{oo}3) \qquad\ \ (\text{single}) \quad\ \ (2\ \text{channels}) \\
&= \textbf{6.5} \times \textbf{10}^{-3} \qquad\qquad\qquad\ \ - \textbf{OK}
\end{aligned}
$$

Fail-safe

$$
\begin{aligned}
\text{probability} &= (P_S)_{\text{HITI}} + (P_S)_{\text{HIVE}} + (P_S)_{\text{HISS}} \\
&= 3.4 \times 10^{-3} + 2.8 \times 10^{-3} + 34.4 \times 10^{-3} \\
&= 40.6 \times 10^{-3} \approx \textbf{4.1} \times \textbf{10}^{-2} \qquad - \textbf{OK}
\end{aligned}
$$

Final system has 2 oo 3 voting for HITIs, a single voting element and two shut-down systems in parallel.

Example 7.2: Hazard analysis of a protective system

In a potentially hazardous chemical plant there are five possible fault conditions, each one of which could lead to a dangerous situation. There are five identical trip systems, each protecting against one of the above fault conditions. Each trip system consists of:

three initiators, each with a fail-danger failure rate = 0.2 year^{-1}
one 2 oo 3 voting element, fail-danger failure rate = 0.4×10^{-3} year^{-1}
three shut-down channels, each with a fail-danger failure rate = 0.6 year^{-1}

and is tested at three-week intervals. The demand rates associated with the five fault conditions are:

2×10^{-2}, 1×10^{-2}, 15×10^{-2}, 3×10^{-2}, 1×10^{-2}

(a) Calculate the total hazard rate for the plant.
(b) Explain fully whether or not this is acceptable and indicate what modifications may be necessary.

Solutions

(a) From Equation 7.12

$$\text{FDT} = \frac{1}{(r+1)} \frac{n!}{r!(n-r)!} F_D^r, \quad r = n - m + 1$$

Initiation

$$2 \text{ oo } 3 - n = 3, \quad m = 2, \quad r = 2, \quad \text{FDT} = \frac{1}{3} \frac{3!}{2!1!} F_D^2 = F_D^2$$

$$F_D \approx \lambda_D T, \quad T = 3/52, \quad \lambda_D = 0.2$$

$$\text{FDT} = (\lambda_D T)^2 = (0.2)^2 (3/52)^2 = \mathbf{1.33 \times 10^{-4}}$$

Voting

$$1 \text{ oo } 1 - n = 1, \quad m = 1, \quad r = 1, \quad \text{FDT} = \frac{1}{2} \frac{1!}{1!1!} F_D = \frac{1}{2} F_D$$

$$\text{FDT} = \tfrac{1}{2}(\lambda_D T) = \tfrac{1}{2} \times 0.4 \times 10^{-3} \times 3/52 = \mathbf{1.15 \times 10^{-5}}$$

Shut down

$$1 \text{ oo } 3 - n = 3, \quad m = 1, \quad r = 3, \quad \text{FDT} = \frac{1}{4} \frac{3!}{3!0!} F_D^3 = \frac{1}{4} F_D^3$$

$$\text{FDT} = \tfrac{1}{4}(\lambda_D T)^3 = \tfrac{1}{4}(0.6 \times 3/52)^3 = \mathbf{1.04 \times 10^{-5}}$$

Total FDT for each system

$$(\text{FDT})_i = (\text{FDT})_{\text{Init}} + (\text{FDT})_{\text{Voting}} + (\text{FDT})_{\text{S/D}}$$
$$= (13.3 + 1.15 + 1.04) \times 10^{-5} = \mathbf{15.5 \times 10^{-5}}$$

Total hazard rate

$$\sum_{i=1}^{i=5} H_i = \sum_{i=1}^{i=5} (\text{DR})_i \times (\text{FDT})_i$$

$$= [2 \times 10^{-2} + 1 \times 10^{-2} + 15 \times 10^{-2} + 3 \times 10^{-2} + 1 \times 10^{-2}] 15.5 \times 10^{-5}$$
$$= 22 \times 10^{-2} \times 15.5 \times 10^{-5} = \mathbf{3.41 \times 10^{-5}}$$

(b) Average background risk in UK chemical industry $\approx 3 \times 10^{-4}$ per year. For 'super' hazardous plant, the **additional risk** should be no greater than $\tfrac{1}{10}$th of this, *i.e.* 3×10^{-5} per year. The target hazard rate for the protective system should therefore be 3×10^{-5}; the achieved hazard is greater than this and should be reduced. The FDT for the trip system associated with the fault condition having the highest demand rate (15×10^{-2}) should be reduced. Since initiation is the highest contributor to FDT, initiation FDT should be reduced for *that* system. If the failure rate of the initiators in that system is halved to 0.1, initiation FDT is reduced to $13.3/4 \times 10^{-5}$ and system FDT is now 5.52×10^{-5}. This reduces the total hazard rate by 1.5×10^{-5} down to 1.9×10^{-5} which is less than the target figure.

Self-assessment questions

7.1 A protective system includes trip initiators which have a maximum fail-safe/fail-danger probability of 5×10^{-2}. Four initiators provide the input signals to an *m* out of 4 voting element which has a maximum fail-safe/fail-danger of 5×10^{-4}.

(a) Discuss fully the advantages and disadvantages of 1 oo 4, 2 oo 4 and 3 oo 4 voting in the above system. Support your answer with relevant detailed calculations.

(b) The initiators and voting element defined above are to be used with a shut-down system, each channel of which has a maximum fail-safe/fail-danger probability of 10^{-2}. Define an optimum overall protective system for this application, giving full justification for your choice.

7.2

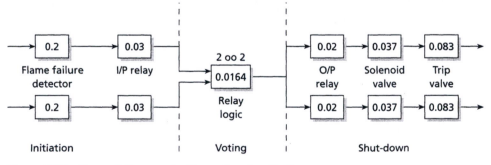

Figure Q7.2   Flame failure protective system in a boiler

A boiler has a flame failure protective system which consists of (Figure Q7.2):

Two initiation channels – each channel consists of a flame failure detector and an input relay.

A single voting element using relay logic – this performs 2 oo 2 voting on the initiation channels.

Two shut-down channels – each channel consists of an output relay, a solenoid valve and a trip valve.

Annual failure rates for each element are as shown.

The target fractional dead time for the system is 0.05; what should the trip testing interval be?

7.3 (a) Explain why a figure of 3×10^{-5} is chosen as the target hazard rate for a potentially hazardous chemical plant.

(b) In such a plant there are five possible fault conditions, each one of which could lead to a dangerous situation. There are five identical trip systems each protecting against one of the above fault conditions. Each trip system consists of:

three initiators, each with a fail-danger failure rate = 0.2 yr^{-1};
one 2 oo 3 voting element, fail-danger failure rate = 0.4×10^{-3} yr^{-1};
two shut-down channels each with a fail-danger failure rate = 0.6 yr^{-1};

The demand rate associated with each fault condition is 3×10^{-2}.

Calculate the required trip testing interval so that the plant meets the above target hazard rate.

Fractional dead time for m out of n voting:

$$\text{FDT} = \left(\frac{1}{r+1} \right) \left(\frac{n!}{r!(n-r)!} \right) F_D^r, \text{ where}$$

$r = n - m + 1$, F_D = single channel fail-danger probability.

Appendix A: Derivation of the reliability of a standby system – cumulative Poisson distribution

Figure 3.8 shows a general standby system consisting of n identical units, each with failure rate λ; the switching system S is assumed to have perfect reliability. Figure A.1 shows that unit 1 fails at time ξ_1, unit 2 fails at time ξ_2, unit i at time ξ_i etc. This means that unit 2 switches in at time ξ_1, unit 3 switches in at time ξ_2, unit $i + 1$ at time ξ_i etc.

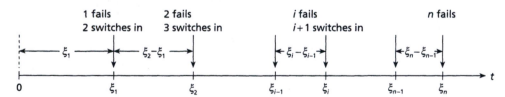

Figure A.1   Unit failure

From Section 2.5:
The probability that unit 1 fails during time interval ξ_1 to $\xi_1 + \Delta\xi_1$ is

$$\Delta F_1 = R(\xi_1)\lambda\Delta\xi_1$$

The probability that unit 2 fails during time interval ξ_2 to $\xi_2 + \Delta\xi_2$ is

$$\Delta F_2 = R(\xi_2 - \xi_1)\lambda\Delta\xi_2$$

The probability that unit i fails during time interval ξ_i to $\xi_i + \Delta\xi_i$ is

$$\Delta F_i = R(\xi_i - \xi_{i-1})\lambda\Delta\xi_i$$

The probability that unit n fails during time interval ξ_n to $\xi_n + \Delta\xi_n$ is

$$\Delta F_n = R(\xi_n - \xi_{n-1})\lambda\Delta\xi_n$$

The probability that all units have failed at times $\xi_1, \xi_2, \ldots, \xi_i \ldots, \xi_n$ is:

$$
\begin{aligned}
\Delta F &= \Delta F_1 \Delta F_2 \ldots \Delta F_i \ldots \Delta F_n \\
&= \lambda^n \{ R(\xi_1) R(\xi_2 - \xi_1) \ldots R(\xi_i - \xi_{i-1}) \ldots R(\xi_n - \xi_{n-1}) \} \Delta \xi_1 \Delta \xi_2 \\
&\quad \ldots \Delta \xi_i \ldots \Delta \xi_n
\end{aligned}
\tag{A.1}
$$

Since

$$
\begin{aligned}
R(\xi_1) &= e^{-\lambda \xi_1} \\
R(\xi_2 - \xi_1) &= e^{-\lambda(\xi_2 - \xi_1)} \\
R(\xi_i - \xi_{i-1}) &= e^{-\lambda(\xi_i - \xi_{i-1})} \\
R(\xi_n - \xi_{n-1}) &= e^{-\lambda(\xi_n - \xi_{n-1})} \\
\Delta F &= \lambda^n \{ e^{-\lambda \xi_1} e^{-\lambda(\xi_2 - \xi_1)} \ldots e^{-\lambda(\xi_i - \xi_{i-1})} \ldots e^{-\lambda(\xi_n - \xi_{n-1})} \} \Delta \xi_1 \Delta \xi_2 \ldots \Delta \xi_i \ldots \Delta \xi_n \\
\Delta F &= \lambda^n e^{-\lambda \xi_n} \Delta \xi_1 \Delta \xi_2 \ldots \Delta \xi_i \ldots \Delta \xi_n
\end{aligned}
\tag{A.2}
$$

Hence the probability that $(n - 1)$ units fail at any time between 0 and ξ_n and the nth unit fails at time ξ_n is:

$$
F = \lambda^n e^{-\lambda \xi_n} \left[\int_0^{\xi_2} \Delta \xi_1 \int_0^{\xi_3} \Delta \xi_2 \ldots \int_0^{\xi_i} \Delta \xi_{i-1} \ldots \int_0^{\xi_n} \Delta \xi_{n-1} \right] \Delta \xi_n
$$

$$
F = \lambda^n e^{-\lambda \xi_n} \frac{\xi_n^{n-1}}{(n-1)!} \Delta \xi_n
\tag{A.3}
$$

Hence the probability that all units fail in sequence between 0 and t is:

$$
F(t) = \int_0^t \frac{\lambda^n e^{-\lambda \xi_n} \xi_n^{n-1}}{(n-1)!} d\xi_n
$$

$$
F(t) = 1 - \exp(-\lambda t) \sum_{k=0}^{n-1} \frac{(\lambda t)^k}{k!}
\tag{A.4}
$$

Thus the system reliability is given by:

$$
R(t) = 1 - F(t)
$$

$$
\boxed{R(t) = \exp(-\lambda t) \sum_{k=0}^{n-1} \frac{(\lambda t)^k}{k!}}
$$

cumulative Poisson distribution

$$\text{(A. 5)}$$
$$\text{(Equation 3.17)}$$

Appendix B: Calculation of P_D, P_S and FDT for systems with m oo n voting

Consider the jth term in the binomial expansion of $(F + R)^n$, where F and R are the single channel reliability and unreliability respectively, and n is the total number of channels. This is equal to (Section 1.4.6):

$$^nC_j F^j R^{n-j} \tag{B.1}$$

and is the probability that j channels fail, *i.e.* $(n - j)$ channels survive.

B.1 | Overall fail-danger probability P_D

If the number of survivors $(n - j)$ is equal to or greater than m then there are sufficient to trip the plant. This means that the system cannot fail dangerously if $(n - j) \geq m$, *i.e.* if $j \leq n - m$. However, if $n - j < m$, *i.e.* if $j > n - m$ ($j = n - m + 1, \ldots, n$), there are insufficient survivors to trip the plant and the system can fail dangerously. The overall fail-danger probability P_D is therefore:

$$P_D = \sum_{j=r}^{j=n} {}^nC_j F_D^j R^{n-j} \tag{B.2}$$

where $r = n - m + 1$ and F_D = single channel fail-danger probability. In a protective system, R will be very close to 1, F_D and higher powers of F_D will be a lot less than 1. This gives the approximate result:

$$P_D \approx \sum_{j=r}^{j=n} {}^nC_j F_D^j = F_D^r({}^nC_r + {}^nC_{r+1}F_D + {}^nC_{r+2}F_D^2 + \ldots + \text{higher order terms})$$

i.e.

$$P_D \approx {}^nC_r F_D^r \tag{B.3} \text{ (Equation 7.10)}$$

B.2 Overall fail-safe probability P_S

If the number of fail-safe failures j is less than m, $i.e.$ $j < m$, then the plant will not trip and an overall fail-safe failure does not occur. However, if $j = m$, m fail-safe channel failures are sufficient to trip the plant and an overall fail-safe failure results. Using Equation B.1, the probability of this occurring is: $^nC_mF_S^mR^{n-m}$, where F_S is the single channel fail-safe probability. The situation where $j > m$ is irrelevant because the plant is tripped by only m channels. Since $R \approx 1$ the overall fail-safe probability is approximately:

$$P_S \approx {}^nC_mF_S^m \qquad\qquad \text{(B.4) (Equation 7.11)}$$

B.3 Fractional dead time FDT

From Equations 7.4 and 7.10:

$$\text{FDT} = \frac{1}{T}\int_0^T P_D(t)\,dt \qquad\qquad \text{(B.5)}$$

and:

$$P_D = {}^nC_rF_D^r$$

where:

$$F_D = 1 - e{-\lambda_D t} \approx \lambda_D t$$

at any time t $(0 \leqslant t \leqslant T)$, provided $\lambda_D T \ll 1$. Thus:

$$\text{FDT} = \frac{1}{T}\int_0^T {}^nC_r(\lambda_D t)^r\,dt$$

$$= {}^nC_r(\lambda_D)^r\frac{1}{T}\int_0^T t^r\,dt$$

$$= {}^nC_r(\lambda_D)^r\frac{1}{T}\left[\frac{1}{r+1}t^{r+1}\right]_0^T = {}^nC_r(\lambda_D)^r\frac{1}{r+1}T^r$$

$$= \frac{1}{r+1}{}^nC_r(\lambda_D T)^r = \frac{1}{r+1}{}^nC_rF_D^r \qquad\qquad \text{(B.6) (Equation 7.12)}$$

where $F_D = \lambda_D T$ is the maximum single channel fail-danger failure probability.

Index